The Butcher's Guide

An INSIDER'S VIEW to Buy the BEST MEAT and $AVE MONEY

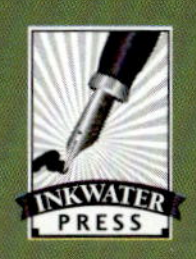

JIMMY KERSTEIN

Scan this QR Code to learn more about this title

Photos by Jimmy Kerstein
Interior Illustrations by Marianne Filbert
Images (bigstockphoto.com): Beef Stamp Design © DaveH900; Grilled Beef Steak © Anna Subbotina; Old Wood © Maxim Loskutnikov
Cover and interior design by Masha Shubin

Publisher: Inkwater Press | www.inkwaterpress.com

Paperback ISBN-13 978-1-59299-972-9 | ISBN-10 1-59299-972-7

Library of Congress Control Number: 2013943319

Printed in the U.S.A.
All paper is acid free and meets all ANSI standards for archival quality paper.

1 3 5 7 9 10 8 6 4 2

THIS book is dedicated to my mother, Lorene Kerstein. She inspired me with her love of cooking as well as her example as a hard working role model. Her cooking knew no boundaries. Although I grew up in a small town with no ethnic restaurants, we ate a variety of dishes, from Chinese, Italian, Mexican, to New York style bagels.

I also owe a great deal of thanks to my wife, Debbie, for her love, support, and writing expertise in putting this book together. When we were first married, I had trouble putting ten words together as we wrote wedding gift thank you notes. I wrote one note while she wrote thirty. Thankfully, she had the patience to teach me how to better express myself with writing.

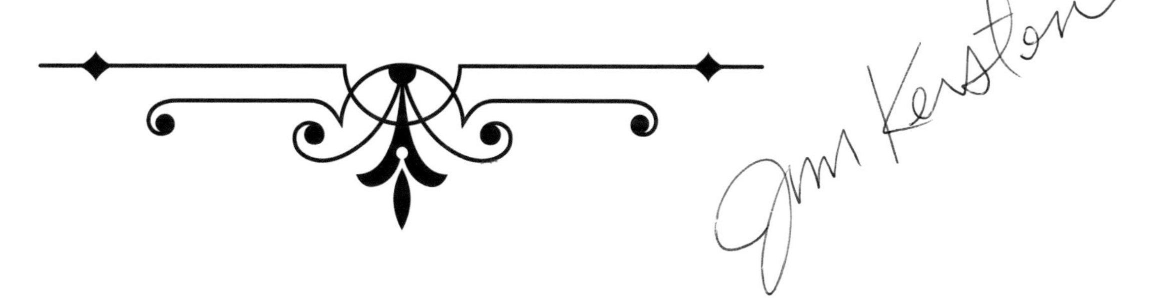

CONTENTS

INTRODUCTION

FOR the forty years that I was in the meat industry, the butcher behind the counter was a tremendous source of information for the consumer. Reading my book, you will understand why that is no longer true.

I truly enjoyed sharing my knowledge with people. I could help them get the right piece of meat for their planned dinner or celebration, and share the preparation method that I knew would make it successful.

My goal for this book is to share that knowledge and empower you to make good, cost effective decisions with your meat dollar. You can link to my website to get more information and share your feedback. I look forward to our interaction.

AN INSIDER'S VIEW TO BUY THE BEST MEAT AND SAVE MONEY

MY name is Jim Kerstein. I have spent close to forty years in the Meat Business. I would like to share with you how to get both the best products and the most value when shopping for fresh meat. My goal is to let you take advantage of my experience in the meat business to eat better and to save money.

As in any business, the people working in the business get the best products for the best price. "Get to know your butcher" has long been a saying in the food industry as a way to buy the best in meat products. Working in the retail meat industry I always enjoyed sharing my knowledge with discerning customers looking for the right cut at the right price.

Unfortunately, the phrase "Get to know your butcher" has less value as the butchers leave the market for the centralized processing plants and the skill level needed to be a butcher diminishes. As the meat business continues to streamline the process, it is harder to find knowledgeable people at store level.

The choices are many in today's marketplace. There is a wide variety of quality and value in meat products, making shopping for meat products very challenging for many shoppers.

Today's consumer has benefited greatly from the changes to the meat industry. Better food safety and a greater selection of meat products are the two major benefits. Enhanced marketing from meat producers and consumer demand for more healthy options, as well as convenience have fueled the expanded selection.

In the following pages I will help to teach you to become your own resource when shopping for the most value as well as the right product for your recipe. With a little help, you will be able to save money and eat better meat. I will help you to understand how to choose the right cut at the right price.

You will learn the basics of good food safety, both at home and in retail. We will cover basic cooking techniques. Most butchers will tell you that there is no such thing as a tough cut of meat as long as it is cooked properly. There are many inexpensive cuts of meat that are wonderfully tender and flavorful when cooked properly. I will share some favorite recipes to help you impress your friends and family with your new found meat knowledge.

AN INSIDER'S VIEW TO BUY THE BEST MEAT AND SAVE MONEY

A BRIEF OVERVIEW OF THE MEAT BUSINESS

Boxed Beef

When I started learning to cut meat many years ago, the skill level needed to become a butcher was much greater than today. Beef was delivered to the stores in front and hind quarters. Lamb and veal were delivered in whole carcasses. Chickens were also delivered whole and then processed into cut-up fryers and parts. The only product delivered as a part of the whole was pork. Demand for hams, bacon and sausage necessitated merchandising the pork differently.

Today's meat world is much different. As labor and shipping costs rose over the years, meat processers further processed the meat before shipping. Packers now trim the excess fat and bone and process the beef into subprimals. A subprimal is a smaller whole section of the beef. Subprimals are vacuum-packed in Cyrovac® (a thick plastic bag) and shipped to the stores in boxes. The commodity name for this vacuum-packed meat sold in boxes is **Boxed Beef.**

Retailers are able buy from the packer an increasing variety of cuts as boxed beef. The packers continue to refine the further processing of the boxed beef, adding to the variety. They also continue to pre-trim the beef of excessive fat and bone. Many of these pre-trimmed boxed beef cuts are a simple slice and tray process to produce case-ready cuts. One great way to save money is to buy subprimal cuts at your local big box store and cut your own steaks and roasts.

Meat Ads

The price of boxed beef cuts rises and falls with supply and demand. Buyers at retail level base their weekly ad specials on lower costs as packers lower prices on excessive commodity inventory. Did you ever notice that within a two to three week period, all of the larger stores around you had the same cut of meat on sale? If the packers are long on chuck rolls because they sell slowly in warm weather, everyone is running chuck steaks on sale at terrific prices. Another way that the larger retailers negotiate lower costs is to place a large order and receive a lower cost.

The front page ad features with the hot prices and the large print are designed to get you into the store. These are usually priced with little or no markup. The smaller print ad products on the inside of the ad flyer are mostly higher margin items to offset the sales of the loss leaders--sometimes suppliers lower their cost so that the retailer can offer an attractive price with a good profit margin.

Retailers refer to people who are shrewd enough to choose only the loss leaders to stock up on as "cherry pickers." As you learn how to select cuts you can choose the best cuts from the display. You can learn to become a cherry picker with a butcher's eye for quality.

Another way to take advantage of ad pricing is to shop late on the last day of the ad. Poorly run meat departments frequently run out of the ad feature on the last day of the ad. Most large chain stores offer "rain checks" to satisfy the customer's needs. You can extend your buying time for lower priced ad product for six weeks or more.

It makes sense to buy and freeze extra of your favorite cuts when you have the chance to buy them at the retailer's wholesale cost or less. The markup or profit margin in fresh meat varies from retailer to retailer. Typically the profit margin on beef ranges from 20% to 40%.

Oddly enough, sometimes the highest priced cuts like tenderloin have the lowest profit margin. Retailers accept lower margins to get more turns or turnover to avoid expensive shrink with unsold cuts.

During the summer months of barbecue time, the higher demand for premium steaks pushes up the wholesale cost. Retailers frequently accept lower margins on premium steaks to be competitive during this time of year.

Profit margins on the rest of the meat category, pork, lamb, veal, and poultry typically are closer to 40%. You might think, Wow, why aren't we all in the retail meat business? Unfortunately, it is not that easy to successfully manage the profit structure of a retail meat business. By the time the costs of ad sales, shrink (dollars lost in the production and sale of meats), supplies, and labor are subtracted from the beginning total profit margin, retailers are left with a profit margin in single digits. Unexpected meat loss, excessive ad sales, or uncontrolled labor costs can quickly turn a profit into a loss.

Locomotion and Support Muscles

Yesterday's butcher had to understand the skeletal structure of the beef to do his own further processing. Butchers in today's meat market remove the meat from Cryovac®-packaged boxed beef and simply slice it into retail cuts. Many of today's butchers would struggle to explain where the cuts of beef come from the whole beef.

Understanding where the cut comes from is an important key to understanding how to cook it properly. There are basically two types of muscles in an animal; locomotion and support muscles. The muscles of locomotion are used for moving. These areas develop more connective tissue as they are used.

When cooking cuts that have more connective tissue, you need to cook with slow moist heat. A chuck roast is an example of a cut that is delicious when slow cooked with moist heat.

Support muscles develop less connective tissue because they are used less. Cooking these cuts is just the opposite of the locomotion muscles. Successful cooking means using hot dry heat. Nothing is better than a great steak grilled over a nice hot grill.

Centralized Cutting Plants

Another change to the meat industry in recent years has been centralized cutting plants. As a way to control costs and be more competitive, many companies are processing their cuts in a centralized plant and then shipping the precut product to the stores for sale to the consumer. The benefits to the consumer are products that have been produced in a controlled environment for better food safety control with a lower cost.

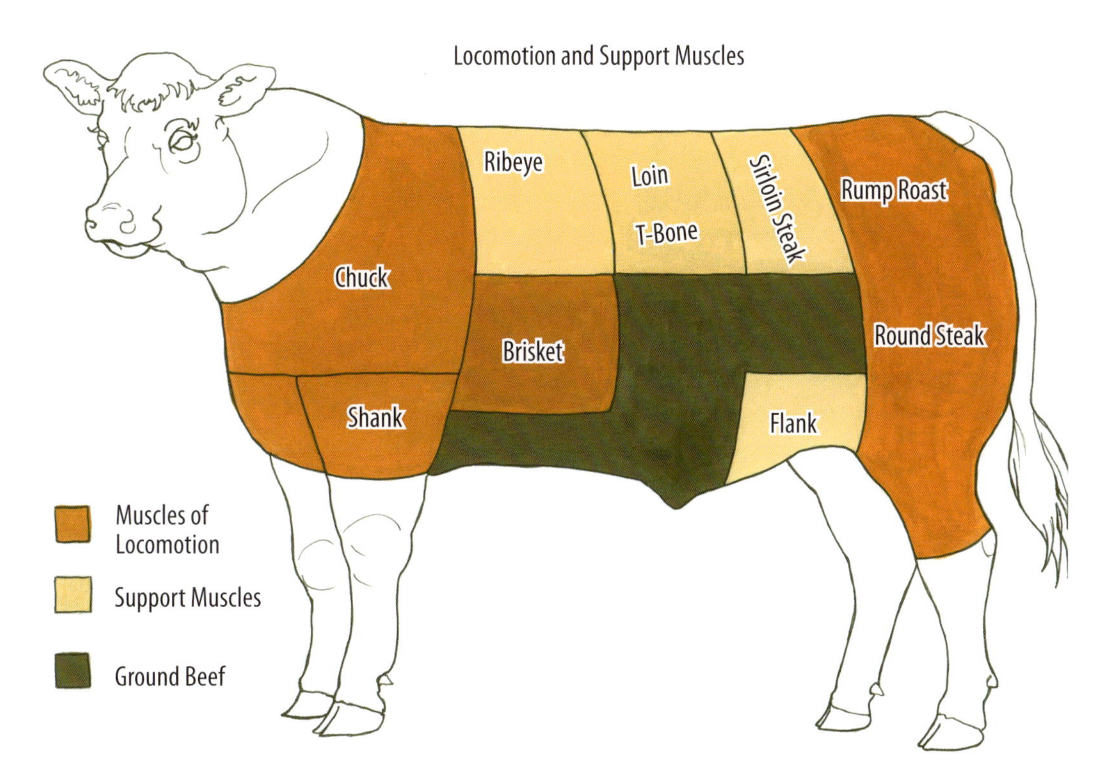

A good example of readily available centralized cut product is chicken. Most stores sell prepackaged poultry as an everyday item.

When I first started in the meat business, we cut up all of our chicken in the store. In our own loosely followed food safety program, we cut the chicken only after we were done processing all of the beef, pork, lamb, and veal for the day. Honestly, I think that the only reason we did it last was because the "real" more skilled butchers came in early and left early. This left cutters lower on the totem pole and apprentices learning the trade to cut up the chickens and to clean up the shop for the next day.

We processed chickens into simple cuts. Bagged whole fryers, cut-up fryers, and the five basic chicken parts: breasts, thighs, drumsticks, wings, and backs/necks. Larger fryers were merchandised as roasting chicken.

Some larger chain stores, like Walmart, fill their entire fresh meat case with meat from centralized processing plants. The products are packaged in modified atmosphere packaging. The industry term for this packaging process is called MAP technology.

MAP products usually come in a deep container with a clear seal of plastic film on the top. Longer shelf life is gained by replacing the oxygen in the package with another gas, such as nitrogen. The retailer benefits by having a longer shelf life to sell the product and the consumer benefits as well with the longer shelf life in their refrigerator.

The shelf life on MAP fresh meats applies only to unopened packages. After opening and exposure to oxygen, the meat's shelf life is similar to other fresh cut meats.

SAFE FOOD HANDLING

SAFE food handling starts where you buy your meat products. You should be able to trust the source of your purchase. Many of the larger chain stores such as Safeway and Kroger have strong company policies and procedures to insure that your purchase is safe. This does not mean that you should only shop larger stores for safe products. It also does not mean that buying only from larger chains insures your safety. Programs and policies are only as strong as the people who execute them. What is important is that you understand what constitutes good food safety.

In a retail environment the retailer is responsible for the safe handling of all of the products offered for sale. It is important that they maintain the proper temperature of the product from receipt in the store to sale to the consumer. Some stores maintain a temperature log of the product as it is received, and keep a log of their cooler and display case temperatures. Bacteria multiply rapidly at temperatures over 40 degrees. Ask the butcher how cold it is in the store's cutting room. Ask how cold their cooler is kept? Many display cases have thermometers on the back in order to monitor the temperature.

In addition to temperature control, it is also important to control the environment in which the product is processed. This means maintaining a clean, sanitary work area to process the products. In some stores the production area is behind closed doors. Without seeing the cutting room you can still get a good idea of the overall cleanliness of the department by just looking around out front. Lift up the front packages to get a look at the bottom of their display case. Does it look like it is cleaned on a regular basis? Even with regular weekly cleaning there may be some dried blood on the bottom in sections where larger cuts, like roasts, may purge. Purge is the natural release of meat juices in fresh meat. It is pretty easy to tell if it is new or accumulated. If the front end of the shop is clean and well kept, chances are the rest is also well maintained.

How does the overall display look? Are the packages all fresh in appearance? Does meat that has lost its "bloom" have an older package or "sell-by" date? If you see meat that appears to be older, with the same date as fresh product, it may have been repackaged and re-dated. This is a policy violation that will cost an employee his job in many retail stores.

"Sell-by" dates

Dating of meat varies across the country. Some use a "**pack date**" as well as a "**suggested use**" or "**freeze-by date**". Some use a "**use or freeze-by**" date only. Others use a date that the meat was packaged. I would question older appearing meats with the same date as fresher looking products.

Rewrapping meat is a necessary process in a meat department. Meat is rewrapped as packages are torn or become wet. Meat is also rewrapped as the prices change with new ads, or to reflect price increases or decreases.

Sometimes careless or unscrupulous people don't maintain the original date on the package. Most modern meat scales update the current date automatically. As the operator scales rewrapped meats, he or she needs to change the function of the scale, changing the date to match the original date on the rewrapped package. Honest people still make mistakes.

If you find questionable meat in the meat case, bring it to the attention of the Department Manager. Most good operators will be happy when you help them to maintain the integrity of their products. If you frequently see "mistakes" on display, consider changing where you shop.

You need to be able to trust that the date on the package is correct. It is your guide to how long you can safely store the meat in your refrigerator. You should be able to store fresh meat in your refrigerator for three to five days from the date of packaging. The meat may lose some of its "bloom" but should not develop any dark spots or have any sour odor.

Freshly cut meat has a nice bright appearance or "bloom" when first cut and packaged. When the freshly cut surface is exposed to oxygen, the myoglobin in the meat makes red meat turn a bright red color. This is much the same process as the way your blood turns bright red as you bleed when you are cut.

One of the most common complaints from unhappy customers is, "How come the center of my ground meat is dark while the outside is bright red?" The outside of the ground beef is red because the myoglobin has been exposed to oxygen.

Most often, the center will turn a more red color if it is exposed to oxygen. The leaner the ground beef, the more likely it is to be dark on the inside. Ground meats with a higher fat content may not be as dark in the center of the package. The fat separates the leaner ground pieces of meat, allowing oxygen to react with the myoglobin.

Ground meats should keep for two to three days. Again, it may lose some of its bloom but it should not have any sour odor. A sour odor could be an indication that the meat used to make the ground beef was not fresh. "Old school" butchers routinely added unsold cuts and yesterday's ground beef to fresh shop trim. The fresh shop trim and the older meats and grind were mixed together to make "fresh" ground beef.

If you do not get the expected storage time in a refrigerator colder than 40 degrees, you should return the meat to the store where you bought it. Repeated returns should result in choosing a new place to shop.

Markdown Meats

Not every package of meat that is cut and offered for sale does sell. How the department handles these cuts is very important. Most stores offer a price reduction on short- coded or older products. Shopping the markdown section is a great way to save money. Markdown meats should be repackaged and frozen if they are not cooked on the day of purchase.

Unfortunately, some stores abuse the selling of markdown meats.

They try to sell products long past their prime. If their markdown meats are dark and tired, don't buy them. Consider looking for a new store to buy meat. If they are pushing the limits of selling unwholesome products they are probably pushing the limits elsewhere.

Shrink

Stores markups or margins have a loss built into them. Fresh meat is a perishable product with a limited shelf life. Responsible retailers understand that they are going to lose a certain percentage of their total sales. This dollar loss is referred to as "shrink".

Some retailers try to control shrink through a variety of unsafe food practices. Historically, ground meats and sausage were used to recycle unsold meats. Unsold cuts were added to the ground beef each day. Unsold ground meats were mixed with each new day's production of ground beef. Unsold pork cuts were processed into sausage. I would hope that the number of retailers still subscribing to these outdated practices is very small.

Food safety programs and policies in many retail chains have addressed this "old school" of managing shrink. Responsible retailers now keep logs of the ground beef production in store. With these logs, stores track the cleaning and sanitation of the grinding equipment, as well as the quantities and products ground. Much of the ground beef produced at store level is made from trim generated from daily cutting.

If the store has need for additional ground beef it is common to grind prepackaged tubes of coarse ground beef. It is important for the store's liability that they are able to track the production chain of all the ground meat products.

Responsible retailers follow the following program. Yesterday's ground beef is reduced and discarded if not sold. Cuts from the meat case are no longer added to the ground beef each day. These cuts are marked down and discarded if not sold. If you do not see markdown meats for sale in your store it would be a good idea to ask what happens to this product.

Cross-contamination

Another critical issue in the in-store production of meat products is cross- contamination. It is important that tools and cutting surfaces are cleaned and sanitized between cutting different species. I honestly don't think that most stores do as good of a job with this as they should. Cleaning and sanitizing slow down the production process. Too many operations are under the gun to produce more products in less time as they try to meet the labor goals.

Proper storage of food products is also a cross-contamination issue. Cooked and ready to eat products should not be stored or displayed under raw products.

The increased awareness of safe food handling has resulted in a safer more wholesome product for today's consumer. Choose the retailer you shop with by looking past the display case. Is it clean? Are there any unpleasant odors? Do the products consistently appear wholesome and fresh? Ask questions. Does the store maintain records of temperature control? Do they maintain grind logs to track sanitation and products ground? What measures do they take to control cross-contamination? Any meat manager that excels at producing wholesome, fresh products should be happy to show you the extra efforts he and his crew make to give you the best.

Food safety has become a more focused issue in the production of meat products. Everyone understands how easy it is to kill a thriving business with a food safety issue. A publicized food safety issue can permanently damage the reputation of even the best food producer. As a group, I trust that most all of the major meat producers are doing their best to make sure the products they produce are food safe.

AT HOME FOOD SAFETY

WHEN you bring your meat purchase home the food safety starts with you. Following good food safety rules is an important part of cooking and enjoying great foods. The USDA stresses four simple steps that are critical to food safety.

Clean – wash hands and surfaces often
Separate – don't cross-contaminate
Cook – cook to proper temperatures
Chill – refrigerate promptly

Food safety experts consider these four steps critical control points to controlling or preventing the bacteria that cause food borne illness.

Cleaning involves more than just keeping your kitchen work surface and tools clean. Ask the question-- does it just look clean or is it sanitary? Clean appearing surfaces and tools are not always sanitary. If you use the same sponge or towel all day to wipe off work surfaces, you may be just spreading around the bacteria growing on that sponge or towel.

A simple mixture of one cap full of unscented chlorine bleach mixed with a quart of water is an effective bacteria killer. Rinse and keep your sponge or towel in this mixture or keep the mixture in a spray bottle to spray on work surfaces and tools. Disposable paper towels are a much safer choice than sponges and reusable cloth towels. Spray tools and let them air dry after cleaning. The bleach and water solution in a spray bottle should be replaced every couple of days for maximum effectiveness. Many home dishwashers have sanitation cycles or sanitize as they clean.

Washing your hands often is an important part of cross-contamination. Wash and sanitize your tools and cutting surfaces as you move between different species of meat as well as cooked, ready-to-eat foods, and raw foods. Designating different cutting boards for cutting raw foods and ready-to-eat foods is a good idea.

The Danger Zone

Refrigeration and temperature control are important keys to food safety in the home. *The danger zone for foods is between 40 degrees and 140 degrees*. Bacteria grow quickly in this range. Most foods are safe for up to a couple of hours in this range as long as the room temperature is not too warm. If you are serving outdoors in warmer temperatures, refrigerate after one hour. It is important to keep cold food cold and to keep hot food hot.

Refrigerate soups and stews in shallow containers to lower the temperature below 40 degrees more quickly. Reheat soups and stews to 165 degrees before serving.

The only way to insure your food is cooked to the proper temperature is to use a thermometer. A good thermometer is your most valuable kitchen tool and will make you a star in the kitchen. An overcooked holiday rib roast is just an expensive pot roast if it is overcooked. Who really enjoys the taste and flavor of dry overcooked or worse yet, undercooked poultry?

Thermometers come in a variety of types and prices. You probably will need more than one--an instant-read thermometer for spot checking current temps and another one that you can leave in the meat while it cooks. Your local kitchen store can help you with your choices.

Thermometer in hand, you need to know the target temperatures for the meats you are cooking. Keep in mind that meats will continue to cook after you remove them from the heat. Depending on the size of the cut, the temperature will increase from 5 to 10 degrees as it rests.

Resting the meat after you remove it from the heat is important. The juices in the meat will redistribute before you cut into it. If you cut into a piece of meat fresh off the heat, much of the juice and flavor will run out.

COOKING TEMPERATURES

Poultry

Poultry cooked to an internal temperature of 165 degrees is considered a safe minimum temperature. At this temperature some dark meats may still be pink. Consumers may choose to cook poultry to higher temperatures because of personal preference. Cooking thighs and drumsticks to 175 to 180 degrees will still result in a moist product with little or no pink color. Cooking breast meat to a temperature higher than 170 degrees will most likely result in a dry product. Inserting an instant-read thermometer into the thickest part of the breast or the thigh (without hitting the bone) is the best way to temp poultry. All ground poultry products need to be cooked to an internal temperature of 165 degrees.

Pork

The USDA recently revised the finished internal temperature for pork to 145 degrees with a three minute resting time. This brings pork to the same standard as beef, lamb, and veal. The three minute resting time is also a new addition. The previous standard for cooked pork was 160 degrees.

Our mothers and grandmothers cooked pork to a much higher temperature because of a fear of trichinosis. Yesterday's commercial pork was raised in a much different environment than today's pork. Their diet sometimes consisted of food garbage that caused the development of trichinosis in the meat.

Thirty years ago the pork sold in stores came in a variety of sizes. There was also an inconsistency in the color of the meat. It would range from a darker red color to the nice white color we associate with today's "other white meat".

Pork for today's market is produced in a much more controlled environment. Better feeding practices and selective breeding have virtually eliminated trichinosis as well as producing a leaner product with a consistent white color.

Leaner cuts from the pork loin such as chops, roasts, and pork tenderloins are often cooked with hot dry heat. Leaner cuts can quickly become dry and tough with overcooking. Many food professionals choose to cook these cuts to doneness with just a touch of pink in the center.

Since trichinosis is killed at 137 degrees, I make the personal choice to pull them from the heat at 140 degrees. The temperature will rise to 145 degrees or more as it rests. Ground pork and raw sausage products need to be cooked to an internal temperature of 160 degrees.

Most hams and smoked pork sausage products are fully cooked when you buy them. If they are fully cooked it will be stated on the label. Most hams and sausages come with reheating suggestions and cooking times on the label. If the smoked product does not say that it is fully cooked, treat it as raw product.

Beef

The suggested food safe temperature for cooked beef is 145 degrees. Ground beef should be cooked to an internal temperature of 160 degrees. Personal taste dictates the finished temperature of more tender cuts such as steaks or more expensive roasts. Many of these cuts are at their juicy best when served rare to medium in doneness.

Finished cooked temperatures for beef (Finished temperatures include the temperature gain during the resting time after the meat is removed from the heat.) Most cuts of meat will gain five degrees or more after the meat is removed from the heat. I target the lower recommended finish temperature.

Rare – 120 to125 degrees – center is bright red with pinkish exterior.
Medium-rare – 130 to 135 degrees – center is very pink.
Medium – 140 to 145 degrees – center is a light pink.
Medium-well – 150 to 155 degrees – center is not pink.
Well-Done – 160 degrees and above – meat is uniformly brown.

Lamb and Veal

The recommended finished cooking temperatures for lamb and veal is 145 degrees. Ground lamb or veal need to be cooked to an internal temperature of 160 degrees. As with beef, consumers sometimes make a personal choice to cook lamb and veal cuts to a lower internal temperature. Use the same cooking methods as you use with beef.

FROZEN MEAT

Freezing meat is a long-term solution to storage. Most of today's freezers are frost free. The biggest danger to frozen meat is freezer burn caused by the appliance keeping the moisture level low. Most of the packaging that meat comes in is not suitable for freezing. The thin plastic film covering the package is easily punctured. The bags that whole fryers come in are pretty air tight and can be overwrapped with freezer paper.

There are lots of choices for wrapping the meat for freezing. Plastic-lined freezer wrap, freezer bags, and vacuum machine bags are all good choices. The key to longer storage in the freezer is to eliminate as much air around the meat as possible. Remember to date every package you freeze. Rotate your frozen meats so that the older frozen meats are used first.

Well-packaged fresh meat will keep in a freezer for up to a year or more. There is some loss of quality for meat that is frozen for any length of time. There is always moisture loss or purge when the meat thaws. Never thaw meat on the counter, out of refrigeration. If you are like me, my mother always thawed meat on the counter. I survived just fine, but I know better now. The outside temperature of the thawing meat may be in the danger zone of 40 to 140 degrees for more than two hours. If you have minor freezer burn on frozen meats, it can be trimmed off prior to cooking.

If you experience a loss of power in your freezer, **do not open the door**. The contents will keep colder longer if the door is kept shut. If the power outage lasts more than a day or two, you may want to quickly remove a package to check the temperature. Remember, the danger zone is above 40 degrees for more than two hours. If you are unsure of how long the meat has been over 40 degrees, toss it out. Food professionals have long subscribed to the adage "**when in doubt, toss it out**".

Refreezing meat that has been thawed has long been thought of as an unsafe practice. The danger in refreezing meat is bacteria starts to grow if the meat reaches 40 degrees or more. Freezing the meat does not stop the bacteria growth process.

There are always quality issues with refreezing meat because of the additional loss of moisture. However, if the meat does not get into the danger zone above 40 degrees, there are no food safety issues. It is okay to refreeze wholesome meats.

Freezing of smoked hams for cold servings, such as sliced ham for sandwiches is not recommended. The cure or brine is injected into the meat in most of today's hams. As the ham freezes, the water in the ham separates the meat fibers leaving a coarse textured ham after it thaws. Roasting or cooking the ham before serving helps to return the more natural texture of the ham.

Efficient use of your freezer is a great way to take advantage of saving money on your meat purchases. Buy extra of the cuts you like when your store has a big sale. Re-process larger cuts or subprimals for your family's use. Comb the markdown section in your store and freeze good buys. Keep in mind that older markdown meats will not freeze as long as fresh cuts.

Thanksgiving turkeys are a good example of buying at the right price and using your freezer to save money. Turkeys are at their absolute lowest prices of the year during the Thanksgiving shopping season. The turkey that you pay 29 cents a pound for probably cost the retailer 75 cents a pound or more. Turkeys are also packaged in a nice tight wrap that stores well in the freezer. We will cover re-processing and refreezing whole turkeys in the Poultry section of this book.

BASIC COOKING METHODS FOR FRESH MEAT

MOST professional butchers will tell you that there is no such thing as a tough cut of meat as long as it is properly cooked. Understanding where the cut comes from helps to decide how to cook it. It is really very simple. Muscles of locomotion develop more connective tissue and are best cooked with slow moist heat. Support muscles don't develop a lot of connective tissue and are best cooked with fast high heat. As we break down the cuts from each species, I will help you understand the best way to cook it.

The range of simple techniques for cooking includes grilling, broiling, roasting, sautéing, frying, braising, stewing, and poaching. Braising and stewing are mostly used for meats that are best when cooked with slow moist heat. The rest are used with both high, hot heat and low, slow moist heat. Understanding the basic cooking methods will help you select the right recipe after you decide whether to cook it fast and hot or to cook it low and slow.

All meat cooking methods start with the basics.

1. Rest the meat at room temperature prior to cooking. (From 20 to 60 minutes, or longer, depending on the size of the cut). Bringing the meat up to room temperature will even out the rise in temperature during the cooking process. The center of the steak or roast will have a more uniform pink or red color when cooked to a temperature that is less than well-done.

2. Season the meat when you set it out to rest. Salt and other seasonings really only penetrate the surface of the meat. Rubbing the meat with a thin layer of oil will help seasonings stick to the meat. If you are marinating the meat in a liquid, let it rest in the marinade at room temperature prior to cooking.

3. Use a meat thermometer. Insert the thermometer into the thickest part of the meat. Avoid any contact with the bone. The only sure way to judge the temperature

of cooked meat is to use a meat thermometer. If you cook a lot of steaks, you may be able to judge doneness by how the steak yields to pressure when you press on it. Using a thermometer in the correct manner helps to insure success.

Remember to take the meat off of the heat 5 to 10 degrees before it hits the target temperature. Let the meat rest between five and ten minutes after you remove it from the heat. Larger roasts can rest for twenty to thirty minutes.

Loosely tent the meat with aluminum foil as it rests. The juices in the meat return to the meat as it rests. Avoid the temptation to slice of just a taste before the meat is ready to cut. If you slice into the meat too early the flavorful juices will run out.

A benefit of this extra time is that it gives you the opportunity to bring the finished meal together at the same time. I use the extra time to toss the salad, to warm the bread or to finish a side dish.

Recipes

I have included several recipes throughout this book. For me, a recipe is just a good outline for a successful dish. I follow the recipe closely the first time I make it. Almost always I make changes to it the second time around.

Understanding the basics of how to cook, and using proper techniques are much more important than following the list of ingredients in any recipe. The good news--the more recipes you try and the more time you spend cooking, the better your techniques become.

Grilling

Grilling is my favorite way to cook meat. There is something very primal about poking at a piece of meat on an open fire. Keep your grill clean and brush it with oil before using it to achieve consistent results.

Grilling includes both the hot, fast method and the low and slow method. Successful grilling sometimes includes both methods on the same cut of meat. Leaving a space on a charcoal grill with no coals underneath allows you the option of indirect cooking, by providing a place to finish thicker cuts that have been seared on the hotter part of the grill.

Searing the meat at the start of the cooking process is important to develop a nice crust to keep in the juices. With the lid on, the grill acts like an oven on the indirect section. You can roast larger cuts of meat or whole fryers by putting a pan in the center of the grill and arranging the coals around the edge.

A nice hot grill with the lid off is my favorite way to grill steaks, chops and burgers that are one inch thick or less. If your grill does not get hot enough, you may have to use the cover to raise the temperature. Be careful--thinner cuts can quickly overcook in a closed grill. I feel that I have a better sense of control over thinner cuts with the grill top off.

Don't move the meat until it has a chance to get a nice sear on the bottom. Use tongs or a spatula to flip the meat. Using a fork will puncture the meat and allow the juices to run out. Be careful not to get carried away with your thermometer and poke too many holes for the same reason.

Gas grills usually have more than one burner. This allows you the option of having more than one heat zone when grilling. It also allows you to start your cooking on high and to then turn down the grill for a more even cooking process. With the lid down and the seared meat placed over indirect heat, your gas grill is a very efficient oven. A small pan with moistened wood chips on the hotter section of your gas grill will give your meats a smoky flavor. Many models even have a thermometer on the outside to monitor the oven temperature.

You can achieve terrific results indoors by using a cast iron pan with raised grooves in combination with your oven for indirect cooking. A cast iron pan without grooves also works well. Some stoves have a built in grill that works well, too. The process is the same. Start with a good sear to lock in the juices and finish cooking thicker cuts and roasts with indirect oven cooking.

Broiling

Broiling can be used on any tender cut of meat that grills or sautés well. Broiling takes advantage of the hot direct cooking element on the top of most ovens. The temperature of most broilers is 450 degrees or more. How do you decide how far from the broiler to cook the meat? That is the key decision. Obviously, thicker cuts should be cooked further from the heat.

I like to keep the oven door slightly open to monitor the cooking process. As with grilling, try to get a nice sear on one side before turning. Try to turn once only during the cooking process. Because of the high heat it is very easy to overcook broiled meats.

The broiler is also very useful to melt cheese or to brown the crust of meats with a bread covering. Many casserole dishes call for finishing the cooked dish under the broiler.

Pan Roasting

Pan roasting is a wonderful way to cook chops and steaks. The meat is first pan- fried in a small amount of oil and then finished to the desired temperature in a hot oven. A cast iron pan works well for this technique. Do not use a nonstick pan for pan roasting. Some nonstick coatings are not safe for higher oven temperatures.

Roasting

The most common question customers ask when roasting meats is, "Should I cover the roast when I put it in the oven?" Most roasts are best when roasted uncovered. Adding a cover will steam the meat as it roasts. If the roast is very lean this can sometimes result in a chewy finished product.

To dry roast, simply season the meat and roast in a roasting pan that will collect the juices. Some roasting pans have a rack in them that holds the meat above the accumulating juices.

Place the meat fat side up to keep the meat juicy. If the roast you are cooking is very lean, you might want to ask the butcher to tie a layer of fat on the top of the roast. The natural fat that is in and around a cut, like a pork shoulder roast, will melt away and keep the meat juicy and tender when slow cooked in a low temperature oven.

There are many schools of thought when it comes to the oven temperature. Many tender cuts benefit from both a shorter, high heat and a lower, slower process. Cooking a prime rib roast is a good example. Some will tell you to start it at a higher temperature to get a good crust on the roast, and to then lower the temperature to finish cooking. Others will tell you to cook it low and slow and then raise the temperature at the end of the cooking process. Honestly, if you let the meat rest before and after roasting, and you use a good meat thermometer to judge doneness, both methods work well.

Braising/Pot Roasting

Braising is another low and slow way to cook tougher meats to mouth watering tenderness. Cast iron pans or Dutch ovens are the preferred choices to braise meat. The heavy metal keeps the temperature even during the cooking process.

Start by browning the meat in a bit of oil on the top of the stove. After the meat is browned, liquid is added and the dish is finished in the oven. Bring the dish to a simmer before moving it to the oven.

You can use a variety of liquids--from wine and stock, to just water--to add moisture and flavor to the dish. The liquid should not cover the meat. You may have to add liquid to some dishes as they simmer. Most braising recipes leave the lid on for all, or most, of the cooking process.

Herbs and aromatics are usually added with the liquid to make a rich sauce as the meat cooks. Cooking times and temperatures will vary with recipes. A tender juicy pot roast "like Mama used to make" is a good example of braised meat.

Sautéing

Sautéing is a quick way to cook more tender thin cuts. Start with a hot pan and add just enough oil to coat the bottom of the pan. The pan should not be covered. Adding a cover to the pan will keep the moisture in the pan. This can cause some cuts to become chewy as they are steamed.

Do not use a nonstick pan if you plan to make a pan sauce. The accumulation of browned bits stuck to the bottom of the pan is called fond. These flavorful browned bits add flavor to the sauce when liquid is added to the pan. Adding liquid to release this flavor is called deglazing. The necessary fond for deglazing will not develop in a nonstick pan.

If you are cooking a larger quantity of meat, cook it in batches. It is important to leave room around each cut of meat so that it browns evenly. Allow each cut to brown before moving it in the pan or flipping it over. If you try to move the meat in the pan too early, the meat will stick to the pan.

Frying

Frying involves cooking more tender meats, usually coated or breaded, in deeper hot oil. Choose oil with a higher smoke point such as peanut oil. Oil temperature is very important for good fried foods. If the oil is not hot enough the end result will be very greasy. Most fried recipes give oil temperature directions. Oil heated to 350 degrees is the norm for most fried foods. Keep in mind that as you add meats to the oil the temperature will drop. Take care to cook in batches and to not overload the fry pan.

Adding the meat to the oil by dropping it away from you will help to avoid nasty burns. Drain the meat on racks or layers of paper towels after cooking. Add additional seasoning if necessary as soon as you remove the meat from the oil.

Stir-fry

Stir-frying is cooking is done in a wok over medium-high to high heat. Use a small amount of high smoke point oil such as peanut oil to keep the meat from sticking to the hot pan.

Cut tender cuts of meat and poultry into bite sized pieces before starting. Having all of your meats and veggies ready to cook before you start is the most important part of stir-frying. Food cooks quickly in a hot wok. Most recipes call for cooking the meat first. Remove the meat just as soon as it is cooked.

Add veggies in the order of their texture, hard veggies such as carrots first, soft bean sprouts last. Add the meat back at the end of the cooking process. Stir-fry is ready to serve just as soon as it is done.

Stewing

Stewing is much like braising except that the meat is covered by liquid after browning in oil. It is also a nice way to cook tougher, more flavorful cuts of meat. Stewed meats are cooked on the top of the stove with a cover and the burner turned to simmer.

As with braised meats, herbs and aromatics are added with the liquid. As the meat slowly simmers, the connective tissue breaks down and becomes very tender. Using recipes to slow cook less expensive cuts by braising or stewing is a great way to stretch your food dollars.

Poaching

Poaching uses seasoned water or oil at a very low temperature. The meat is covered with a liquid and then simmered at a very low temperature for just a short time. The pan is then removed from the heat and covered as the cooking process completes.

Poaching is a quick easy way to cook tender boneless chicken breasts for chicken salads and enchiladas. It is nice to have a low fat cooking option to offset the occasional fried chicken.

Cover the chicken breasts with salted water. Add fresh herbs. Oregano, thyme, tarragon, and rosemary are some of my favorites. Add crushed garlic and whole peppercorns for more depth of flavor.

Bring them to a low simmer for one to two minutes. A simmer is the point just before the liquid boils. Look for a consistent rise of small bubbles around the edge of the pan. Be careful not to let the liquid come to a boil. Boiling the meat can make the chicken tough.

Remove the pan from the heat and let it stand covered for another twenty minutes. The cooked chicken can then be cooled and cut into chunks for salads or be shredded for enchiladas.

SEASONING MEAT

LEARNING how to season meats is one of the true pleasures of cooking and eating meat. The seasoning of most meat dishes starts with simple salt and pepper. In some cases this is all of the seasoning the meat needs to showcase its meaty flavor.

Even salt and pepper is not without choices. Should you salt the meat before, during or after cooking? There are those that will tell you that salting the meat before cooking will draw the moisture out, making the meat dry.

What kind of salt should you use? Supermarkets today offer a variety of salts from seasoned salts, table salts, kosher salt, to exotic salts such as red Hawaiian salt. Even pepper has choices. Should you use canned pepper, (white or black), whole pepper corns, (white, black, or red)? How coarse should you grind the pepper corns?

Salt

I subscribe to the group that says you should salt meat before cooking. I add salt and other seasonings to meat prior to letting it rest at room temperature for cooking. Most of the time I rub a thin coating of olive oil on the meat to help the salt and seasonings stick to the meat. When the meat is grilled or cooked in a hot pan, the salt and other seasonings combine with the natural sugars of the meat to caramelize into a nice brown crust.

I like to use kosher salt for most of my meat cooking. The thin flakes of kosher salt are easier to distribute evenly, and melt with the natural juices more easily.

There are a variety of gourmet salts available. You can choose from smoked salts, natural gray sea salts, fleur de sel, or Hawaiian salt, among many. They each add a unique salty flavor to cooked meats. Be cautious the first time you use a new salt. It is very easy to over salt. I love the salty, mineral taste of red Hawaiian salt on a grilled steak.

Pepper

Fresh cracked black pepper is the best way to go. There is no comparison to pre-ground canned black pepper. The investment in a good pepper grinder is one you will never regret. Choose one that has multiple settings for the coarseness of the grind. I buy Tellicherry peppercorns in a large container at my local big box retailer and add red peppercorns for color. I can't tell you that the red peppercorns add any flavor, but they look nice in the pepper grinder.

Much of the heat in fresh cracked black pepper dissipates during the cooking process over high heat. I usually am very generous when seasoning steaks cooking over high heat. Steak Au Pave is a classic steak dish in which the steaks are coated heavily with fresh cracked black pepper before searing.

Garlic

Next to salt and pepper, garlic is probably the next most popular seasoning for cooking fresh meat. Garlic comes in many forms, from fresh and roasted garlic to granulated and powdered garlic. It is often combined with salt to make a ready-to- use seasoning. I put fresh garlic cloves in a decanter of olive oil to add a touch of garlic when cooking.

Garlic can add a different garlic flavor depending on how it is cut and when it is added to the cooking process. Whole roasted garlic is sweet in flavor. Add whole garlic cloves to the start of the cooking process to add a mild garlic flavor in soups and stews. Sliced garlic will add a more intense flavor to dishes than whole garlic. Adding finely minced garlic to the end of a cooking process will result in a stronger garlic flavor. When adding minced garlic to a sautéed dish, add it last and be careful to not burn it. The bitter burned taste of garlic can ruin your dish.

I prefer granulated garlic to powdered garlic as a dry seasoning. Like Kosher salt, the larger granules are easier to distribute over the meat. The granulated garlic also mixes more evenly in dry rubs.

Herbs

If fresh herbs are available, they are always my first choice. If you have time and space, a small herb garden can be rewarding in many ways. Most supermarket produce sections have fresh herbs for sale. Wrap the fresh herbs in moist paper towels and refrigerate them in plastic bags to extend the shelf life. I sometimes cut the stems of fresh herbs like cilantro and keep them in a glass of water. A bouquet of the fresh herbs used in preparation adds a nice fresh gourmet touch to your kitchen for a company dinner.

A general rule of thumb for substituting dried herbs for fresh is to use half as much. Leafy herbs such as oregano and thyme benefit from crushing in the palms of your hands before adding them to your dish. Crushing the herbs releases the natural oils.

Dried herbs have a short shelf life. Three months is close to the limit for dried leafy herbs. I like to buy dried herbs in the bulk section of my supermarket. They are much less expensive and I can control the amount I buy.

Some herbs are natural companions to cooked meat. My favorite is thyme. I think you can add fresh thyme to almost any meat dish to elevate it to a higher level. Fresh oregano, basil and rosemary are other favorites. Thicker-leafed herbs like thyme and rosemary have more intense flavor. Start with less and add more to taste. Add herbs to the end of the dish for more herb flavor.

Spices

For centuries we have been using spices to add to our enjoyment of meat. Spices were once so popular, they were used as currency. The addition of spices is very basic to our enjoyment of meats. Where would we be without the secret spice and herb recipes of barbecue masters and the famous chicken colonel?

The addition of different spices and spice blends can change the whole complexion of any meat dish. It can change the entire country of origin of the dish. Different combinations of spices can even be traced to specific regions of countries.

Buying ground spices is not unlike buying dried herbs. Buy small amounts in a specialty spice shop or in the bulk section of your local supermarket. The shelf life of ground spices is similar to dried herbs. If you have cans of spices you have not used in the last six months, throw them out and start fresh.

Whole seed spices like fennel, cumin, and celery seed will keep until they are ground or crushed. Toasting whole spices briefly before grinding will help to maximize their flavor. Use a coffee grinder or a mortar and pestle to grind whole spices. Running a small amount of uncooked rice through the coffee grinder is a good way to clean it between spices.

Most spices are used at the start of the cooking process either as a rub or added to the cooking liquid. Their more intense flavor melds with the other seasonings as the meat cooks.

The popular sweet and salty flavor we all enjoy can be enhanced by adding spices to the sugar and salt combination. Adding spices and the flavors of chili powders is the foundation of many secret barbecue rubs. Try a variation of my spice rub recipe on page 110 next time you barbecue.

Marinades

Marinades are a popular way to infuse flavor into meats. The variety of flavors you can add to meat is endless.

The length of time the meat is in the marinade is one of the keys to successful marinated meats. Most recipes state a range of time for their particular recipe. Odd, but you can get a similar result cooking a meat marinated two to four hours as cooking a meat that has been marinated overnight.

Marinating longer than overnight is usually not a good idea. The marinade flavor can overwhelm the meat or the texture of thinner cuts can change. Leaving meats in a marinade containing tenderizers can result in a mushy texture.

Reusing marinades or using leftover used marinade as a sauce is also not a good idea. There are serious food safety issues with using leftover used marinade as a sauce. Reserve a portion of the marinade when it is first made if you would like to reduce it to use as a sauce.

Commercial producers use a method called "tumbling" to add marinades to meat products. The meat is placed in a device called a tumbler. The tumbler shortens the marinating time by creating a vacuum inside. This helps the meat absorb marinade more quickly as the tumbler turns.

A food syringe is a good way to add marinade to the center of larger cuts or whole poultry. This also shortens the marinating time for these cuts.

If you are grilling meats that have been marinated, dry the surface of the cut, and coat the surface with a bit of oil before you put the meat on the grill. The dry surface will sear more easily than a wet one.

Brining Meats

Brining meats is another way to add flavor to meats. Brining also adds moisture to cooked meats. The idea is simple. Meats lose water and moisture during the cooking process. Brining meats before cooking adds water to the precooked meat. The end result is a more moist finished meat dish.

The basic recipe for brine is 1 ½ cups of Kosher salt to one gallon of water. If you use table salt, choose non-iodized salt and reduce the amount to one cup per gallon. Add more flavors with the addition of sugar, peppercorns, and dried berries or herbs. The addition of sugars will help the meat to brown as the sugars caramelize during the cooking process. Taste first before adding salt to the sauces and gravy made from brined meats.

The brine should be simmered to blend the ingredients. Use less water and add ice to cool the brine before adding the meat. The brine should be cool before using. All brined meats should be refrigerated during brining.

Pork and poultry seem to benefit the most from brining. Whole chickens and smaller cuts of pork are usually in the brine for four to six hours. Whole turkeys and larger roasts brine for twelve hours or more.

BUY THE BEST BEEF

BUYING good beef is probably the most difficult choice for a meat consumer. Beef offered for sale at retail stores comes in an overwhelming variety of quality and cuts. Today's consumer can choose from USDA graded beef, USDA inspected and ungraded beef, branded beef, grass-fed beef, natural beef, and mail order beef. The list seems to go on and on.

Understanding the USDA grading system is just part of the knowledge necessary to know what you are buying. Fancy steakhouses across the country make a living buying the best beef. If you know what you are buying you can produce the same result at home for a fraction of the price.

USDA GRADED BEEF

The first thing to understand about the USDA grading system is that not all of the beef sold is graded. If the label of the beef offered for sale does not include the words USDA Prime, USDA Choice, or USDA Select, it is ungraded beef.

There is no law that says that all beef sold must be graded. It costs the producer money to have the USDA inspect and grade the beef. After determining the grade, the beef carcass is stamped with an ink roller. For a variety of reasons some producers choose to not pay the USDA to grade their beef. Ungraded beef is sometimes sold as "no-roll", i.e., no ink stamp was rolled on the beef. Retailers buy this beef at a lower cost and give their own marketing name to it. If a local retailer sells beef without a USDA grade and a fancy sounding name such as "Premier Beef" or "Red Star Beef" it could be ungraded no-roll beef.

All meat producers who produce meat for resale have a USDA inspector on site to help insure a food safe product. One exception is the small farm kill operation selling meat directly to the consumer and not for resale. Most of the beef sold through these outlets is sold directly to the consumer as cut and wrapped sides of beef.

The purpose of the USDA beef quality grades is to evaluate the tenderness, juiciness, and flavor of the meat. The beef is judged by two basic criteria, marbling and maturity.

Marbling is the dispersion of fat within the lean. The best marbling is small flecks of fat evenly spread across very fine textured lean. It is an unpleasant truth that fat equals flavor.

The maturity of the beef is the physiological age, not the chronological age of the beef. Beef graders can judge the age of the beef by the size and color of the bone structure. The ribs of the beef change from a round shaped red bone to a flat white bone as the animal ages. Some cartilage turns to bone in older animals.

USDA Prime Grade Beef

The top grade of beef is USDA Prime. Most of the prime beef is sold to steakhouses. Prime beef is also sold through mail order outlets. You can find prime steaks for sale in upscale food retailers and specialty stores. It is worth the effort to source out a retailer selling prime steaks. If you compare the serving cost of a raw prime steak to the serving cost of an upscale steakhouse, the prices look reasonable. The best thing you can do is to find a retailer who dry-ages the prime beef.

Many consumers have a fear of ruining an expensive cut of meat in the cooking process. The reality is that the higher quality meat helps to insure success. Next time you have the urge to treat yourself to an expensive steak dinner, try cooking a Prime Grade steak at home.

USDA Choice Grade Beef

The most widely marketed grade of beef is USDA Choice. USDA Choice beef includes the range of Choice beef just under USDA Prime to the beef just above the leaner, less tender USDA Select beef.

To separate this range, choice beef is marketed under three USDA quality grades. Top Choice beef has marbling graded as MD, or moderate. Middle Choice has marbling graded as MT, or modest. Bottom Choice has marbling graded as SM, or small.

The USDA Choice label on the beef at retail level does not tell the consumer where the beef fell on the grading scale for choice beef. The level on the grading scale for choice beef is probably more important when purchasing more tender cuts, such as steaks.

Have you noticed an inconsistency in the tenderness and flavor of steaks purchased as USDA Choice? You could be buying beef from both ends of the choice scale. When shopping for Choice graded beef, choose beef with more marbling. The marbling in the beef adds flavor and tenderness.

Some retailers pay a premium price to purchase only top choice beef. Because of this their prices may be slightly higher than the competition. The payback is a more consistent quality product. Some branded beef producers such as the **Certified Angus® Brand** have higher grading standards to insure that their product is the best of the Choice grade.

Certified Angus Beef is a registered trademark to promote their high quality standards and should not be confused with other brand names with the words Angus or Black Angus. Certified Angus Beef® is certainly worth any extra money it may cost. It is my first choice when buying an affordable top choice beef steak or roast.

USDA Select Grade Beef

USDA Select is the leanest of the beef quality grades offered for sale at the retail level. It has a marbling score of SL, or slight marbling. The extra marbling of fat in higher grades of beef is what gives it that extra flavor and tenderness. Select beef sometimes also has a lower meat to bone ratio than the more marbled higher grades.

Select beef is a good choice for consumers who dislike fat in their meat. It can also be a good budget choice. It is usually priced lower than the higher grades of meat. The downside is that it can be less flavorful and less tender.

BRANDED BEEF

Branded beef offers another interesting choice for the consumer. Selling beef as a brand name is a way for beef producers to differentiate their product from the other choices. Producers of organic or natural beef frequently merchandise their products under a brand name.

Branded beef producers set their own parameters for producing a consistent product. It helps consumers to identify the beef that best fits their needs and lifestyle. It also helps the producer develop a loyalty to the brand. Stores that sell branded beef usually have pamphlets available near their display explaining the virtues of their product.

Some retail chain stores attach a brand name to ungraded or select graded beef to give it more selling appeal. The lower cost of this beef allows them to offer a price competitive product. The only way to judge the flavor and tenderness of this beef is to give it a try. You will soon know if you get good value for your money.

Many natural raised and organic beef are sold as branded beef. The USDA maintains standards for natural and organic beef.

GRASS-FED BEEF

Most of the beef produced today spends the last few weeks prior to harvest in a feed lot, eating a diet of grains to increase the marbling in the finished product. Up to that point, the beef are raised in pastures eating a diet of grass. Beef sold as grass-fed do not go through the feed lot stage. They spend their entire life eating only pasture grass. Because of the way it is raised, grass-fed beef is leaner with less internal fat.

Grass-fed beef is a healthy alternative to eating feed lot beef. It has less fat and is higher in some nutrients than feed lot beef. Proponents of grass-fed beef argue that the process of raising grass-fed beef is more humane. They feel that the cattle are more naturally suited to being raised to harvest in pastures only.

The downsides of grass-fed beef are a higher cost and more limited availability. Grass-fed beef takes longer to produce and weighs less at harvest time compared to feed lot beef. This all adds up to a higher cost of product. Some grass-fed beef producers market their beef as a brand name beef. Grass-fed beef producers must meet USDA standards to label their beef as grass-fed.

If you have never tried grass-fed beef, I would encourage you to give it a try sometime. If you eat less beef because of health concerns and the cost is not an issue, it might be the beef for you.

NATURAL AND ORGANIC BEEF

The demand for all natural and organic products has grown exponentially in the last twenty years. Beef and other meat products have followed suit as more consumers make healthy lifestyle choices.

Most producers of natural and organic meat processers follow a "**never ever**" policy. This means that from birth, the beef is never ever given growth hormones or antibiotics. In addition the diet for the beef is pesticide free. Old fashioned free range grazing is an important part of raising natural and organic beef. This does not mean that all natural and organic beef is grass-fed only. Many still include feed lot time to finish the product.

Many natural raised and organic beef are sold as branded beef. The USDA maintains standards for natural and organic beef. Producers of natural, organic, and grass-fed beef must meet the minimum USDA standards to label their product as such. Many producers choose to exceed the USDA standards. You can usually pick up POS (point of sale) material near the display to learn the benefits of each product.

The cost of raising natural and organic beef is higher than the cost of the mainstream beef producers. This higher cost translates to higher retail prices. The good news is that many consumers choosing naturally raised product are not as price sensitive.

Choosing natural or organic beef is a personal choice. Many people closely monitor the foods that they eat. I think it is terrific that we have that choice. I have eaten some very flavorful natural and organic beef. However, for me, the biggest determining factor for choosing beef is flavor.

AGING BEEF

Aging is another way to improve the flavor and tenderness of the beef. There are basically two ways to age beef. They are dry-aging, and wet-aging. Because of the increased cost of production, dry-aged beef is usually only available in upscale gourmet retailers or specialty food stores. Some finer steakhouses also dry age beef in-house. Most of the beef sold at retail as aged beef is wet-aged beef.

Dry-aged beef is hung in a cooler for fourteen to twenty-one days or more. The process involves controlling the cooler temperature and humidity while allowing the air to circulate around the beef. As the beef dry ages, it develops a stronger beef flavor as well as additional tenderness. Because the lean in the beef is 70% water, the beef also loses weight in water loss the longer it ages. This loss can be 15% to 20% of the weight of the beef. The beef also develops a dark hard covering that must be trimmed, adding to the cost of the finished product. Another downside for retail sale is the reduced shelf life of aged beef. If you can find it, and can afford it, dry-aged beef is certainly worth a taste test.

Wet-aging replaced dry-aging as producers started packing the beef in Cryovac®. Cryovac® is the process where the beef subprimals are vacuum-packed in thick plastic bags. The Cryovac® beef is then boxed and shipped to stores for further retail processing. "Boxed beef" is the industry standard of mass production.

The beef stored in Cryovac® ages with time as well. Most wet-aged beef is aged a minimum of fourteen days. As the beef ages, it purges, or loses moisture. The weight loss is not nearly as dramatic as the weight loss in dry-aged beef. Beef that is wet-aged is also milder in flavor than dry-aged beef. A period of wet-aging does add to the tenderness of the beef.

TENDERIZED BEEF

Another way that beef is tenderized is through the use of a mechanical tenderizer. This process is called "needling". In this process a series of thin needle-like blades pierce the meat, cutting through the connective tissue. Many steakhouses and restaurants serve needled steaks to insure a tender finished product. The brand name of the machine most often used is **Jaccard**®.

Portion control restaurant suppliers and some large box stores run the steak subprimals through a Jaccard® machine prior to cutting into steaks. The good news is you can buy a hand held model for $40 or less. Check with your local kitchen store or search Jaccard® online. Adding a Jaccard® to your kitchen tool box is a worthwhile investment.

BUYING SIDES OR QUARTERS OF BEEF

As a rule, I don't recommend buying sides or quarters of beef. There are a couple of downsides to buying a side of beef. You can save more money shopping weekly meat specials. You also save money by doing your own portion control when you package the meat for freezing.

The only real plus to buying a side of beef is if you have a good relationship with the producer and feel that the quality of beef is superior and worth any extra money. You could also say that there is a convenience factor to buying several months worth of beef as one purchase.

Sides and quarters of beef are sold by "hanging weight". This means that you pay for all of the trim and waste lost in the cutting process. You can expect to lose around 25% of the total weight as a cutting loss if the beef is processed into mostly bone-in cuts. Expect more loss if you ask for more boneless cuts. A side of beef costing $2.25 per pound will actually cost you $3.00 a pound after cutting loss. The cost of cutting and wrapping is sometimes listed as a separate cost. You need to add the cutting cost to the cost of the beef to get the total cost.

The lower average cost of a side of beef may sound pretty good if you think in terms of premium steaks like rib-eyes and T-bones. Keep in mind that the total yield of these two steaks will be 10% or less of the total hanging weight. The average cost will look much higher if you think about all of the ground beef, chuck roast, round steak, and soup bones that you are buying. Compare these costs to the weekly ad costs at your local supermarket.

Shopping the weekly ad specials enables you to choose the cuts you like best. When you buy a side of beef you buy all of the cuts. Every subprimal includes less desirable and more desirable cuts. If you like your T-bone steaks with the large tenderloin you can choose only those cuts from the ad display. When you buy the whole side of beef, you get all of the T-bone steaks. You get the steaks with little or no tenderloin as well.

Asking a locker beef producer to do your family's portion control costs you money. Cut and wrapped for a family of four is up to their discretion. Most often you will discover that you have more than enough for a dinner for four. Repackaging store bought meat will enable you to do your own portion control. **Buy only the cuts you like, do your own portion control, and save more money.**

SAVE MONEY CUTTING YOUR OWN BEEF

One way to save money buying beef is to cut your own. Many of the boneless subprimal cuts available are easy to slice into steaks and roasts. You will need a sharp knife and proper packaging materials to cut steaks and roasts for your freezer. The cutting of some subprimals also generates trim that can be made into ground beef. A small meat grinder will be needed to grind the trim. Freeze any meats that you do not plan to use in three to five days.

Knives/steels

My recommendation for the "must have" home butcher's knife is a six inch, semi- flex, boning knife made by Victorinox. This knife has long been a favorite of many professional butchers. It comes in both a straight and curved blade. I prefer the curved blade. The flex in the blade makes it easier to move the knife around odd shaped bones when removing them.

Victorinox knives are readily available. Expect to pay between $20 and $30 for a boning knife. This high quality knife will soon become a favorite and will last you a lifetime.

In addition to the boning knife you will need a larger slicing knife. Choose a knife that is eight to ten inches long. Most beginners find that the smaller eight inch size is easier to use. A good quality chef knife works just fine for most home butchers.

An electric slicing knife is also a reasonable choice for slicing thicker cuts or steaks.

Many kitchen stores carry a line of Victorinox knives. If you type in "butcher knives" on the search engine of your computer you will soon find a source to buy one. Many large metropolitan areas have a butcher's supply store that stocks them at a very reasonable price.

Keeping a good knife sharp is important to enjoying its use. The easiest way to accomplish this is to have your knives professionally sharpened. Well-stocked kitchen stores carry a variety of mechanical knife sharpeners that will help you to keep a sharp edge on your knives.

To keep a sharpened blade straight you will also need a steel. Some folks think that the steel is a sharpening tool. Most steels do little to sharpen the knife. Their use is in keeping the sharp edge of the knife straight.

As you use the knife the sharp edge will roll and the knife will dull. The cutting edge of a well-sharpened knife is very thin. If you look lengthwise at the edge of the blade you will see that the knife tapers to a slender "V". Proper use of a good steel will help to keep the edge in a sharper "V" shape.

Wearing a cut resistant glove is not a license to be careless when using a sharp knife.

The safest way for most beginners to use a steel to correct the edge is to hold the steel vertically, with the point of the steel on the cutting board. Place the knife at a right angle at the top of the steel, with the point of the knife away from you and the butt of the knife touching the steel.

Tilt the blade slightly and keeping the blade horizontal, slide the knife down the steel and pull the blade toward you. Repeat this process three to four times and switch the knife to the other side of the blade. Steel a like number of times on both sides. With practice you will develop a rhythm--a consistent even pull across the steel.

Like knives, steels come in a variety of shapes, sizes, and qualities. Most meat professionals prefer a smooth hard steel ten to twelve inches long. Ceramic is an economical choice for a steel; however it will break if dropped.

The only steel that actually sharpens a knife is a diamond steel. Diamond steels have tiny diamonds embedded in the surface of the steel. The diamond steel can be a useful tool to re-sharpen the knife between regular sharpening. Using a ceramic steel after using a diamond steel will help to smooth the rough edge left by the diamond steel.

Cut resistant gloves

If you have a sharp knife in one hand, you should have a cut resistant glove on the other hand. Many large chain stores have made the wearing of cut resistant gloves mandatory for their butchers. If the people who cut meat all day long wear one, you certainly should. Cutting gloves used to be made out of metal links. They were awkward and clumsy to wear. The newer gloves are made out of Kevlar® or similar fabric and are much more flexible and comfortable.

Wearing a cut resistant glove is not a license to be careless when using a sharp knife. You should always be careful when you have a sharp knife in your hand. The glove will save your hand from all of the cuts and nicks associated with learning how to use a knife.

You can purchase a glove and most of the same places you buy your knife. The cost will be $15 to$20. Some sporting goods stores sell fish filleting gloves at a good price.

Meat Grinders

Spending $30 or $40 on a good knife and a cutting glove seems pretty reasonable. Spending $200 to $600 for a home meat grinder is a commitment most of us are not willing to make. If you own such a grinder you probably do not need my book.

Fortunately, there are several less expensive choices to use to grind your own meat at home. I use the grinder attachment for my large mixer. My mom used to use a small hand crank grinder that she clamped on the kitchen table. You can still buy similar grinders at most kitchen stores.

A food processer is a good option to grind small amounts of meat. Cut the meat into one inch cubes and chill in the freezer for 20 to 30 minutes before you put it in the food processer. You want the meat to be just slightly firm, not frozen. Do it in small batches of half pound or less.

Chilling the meat in the freezer for a short time is a good idea any way you grind it. Store bought ground beef does not compare to any ground meat you make at home. You control the fat content and know exactly what cuts are in your ground beef.

PACKAGING MEAT FOR THE FREEZER

Getting a good airtight wrap on your newly cut meat treasure will make certain you enjoy it to its fullest. The biggest danger to frozen meat is freezer burn. Most new freezers control the buildup of ice by removing the moisture in the air. Any loose wrap or hole in your packages will result in freezer burn.

Home vacuum-packaging machines offer the most airtight package. If you can afford one, it is certainly the way to go. They are useful for a wide variety of food product storage.

Plastic-coated paper freezer wrap is the next best choice. Double wrapping with either plastic wrap or a second wrap of paper wrap is the best way to go.

Turn the square of the paper with a corner facing you and the opposite corner away from you. Place the meat in the middle and fold the corner near you toward the other corner. Tightly fold in the outside corners and roll the package toward the opposite corner.

Tape and double wrap in the same manner if you expect to freeze the meat for more than a month or so. If you have ever made homemade egg rolls or burritos the technique is much the same. Don't forget to date the meat when you label it.

BUY BIG AND SAVE

Many large public wholesale stores offer primals just as the stores buy them. Some retail stores sell whole primals as well. You might have to look around a bit to find the right combination of cost and quality that will meet your needs. If you choose a leaner, lower grade of beef consider using a Jaccard® to tenderize it prior to cutting.

There is waste of fat and purge with each primal. The primals in the following chart are all closely trimmed. The fat covering is ¼ to ½ of an inch. If you buy primals with a thicker fat covering, expect a lower yield. You can judge the fat covering by looking at the end of the primal. The chart will help you understand how much meat you can expect to get from primals you might cut at home.

The third column lists the expected average yield percent for each primal. The yield is the amount of usable meat from each primal.

To know the true cost of the meat, divide the cost per pound of the meat by the yield percent. **Example; Beef Strip Loin, cost $5.60 per pound/divided by 77% equals a steak cost of $7.27.** There is no ground beef trim generated in the cutting of a strip loin.

Keep in mind that the yield weight also includes ground beef weight. This means that you are paying the same price per pound for ground beef as the steak or roast cuts in primals that generate ground beef trim. The small amount of ground tenderloin is probably worth the cost of tenderloin. Fresh ground chuck is certainly worth the same cost as chuck roast or steak.

Cutting meat across the grain is an important part of producing a tender cut of meat. The fibers in fresh meat run in a common direction. Cutting the fibers across the grain makes the meat tender.

PSMO Beef Tenderloin – Yield 88%

The PSMO tenderloin is the entire tenderloin of the beef. Because it includes both the butt end and the smaller tail of tenderloin, it has more waste. The small end of the tenderloin is too small in diameter to cut into steaks so is included as ground beef. You can choose to cut it into smaller tenderloin tips that are delicious sliced thin and sautéed. A quick pan sauce is all that is needed to make a terrific pasta dish.

There are a couple of ways to slice PSMO tenderloin into steaks. The first is the easiest. The grain of the tenderloin runs end to end. Simply start at the large end and slice across the grain. Cut a thin slice of ½ inch or less to square up the end of the tenderloin. Add this thin slice to the tail piece for sauté or ground beef.

Before slicing, slide your knife under the shiny silverskin and remove it. Tip the cutting edge of the knife up and slide it along between the meat and the silverskin. You will probably have to do this in sections.

PRIMAL	*Avg Wt.in bag*	*Avg Yield %*	*Avg wt yield*	*Beef Cuts*	*Quantity*	*Cut Thickness Size*	*Ground Beef*
PSMO Beef Tenderloin	6.25	**88%**	5.50	Tenderloin Steak	9 Steaks	1 ¼ inch	½ lb
Beef Butt Tenderloin	3.00	**95%**	2.85	Tenderloin Steak	5 Steaks	1 ¼ inch	none
Beef Strip Loin (New York)	13.00	**77%**	10.01	New York Strip Steak	14-16 Steaks	one inch	none
Beef Rib-eye Boneless	15.00	**76%**	11.40	Rib-eye Steak	16-18 Steaks	one inch	none
Beef Top Sirloin Butt Peeled	10.00	**92%**	9.20	Top Sirloin Steak	12-14 steaks	1 ¼ inch	½ lb
Beef inside Round Peeled	17.00	**93%**	15.81	Top Round Steak	10 steaks	1 ¼ inch	1.5 lbs

The silverskin shrinks when the steak cooks and will cause the steak to curl. Removing it will make the steaks much more enjoyable.

As the steaks start to get smaller as you cut further down the tail, butterfly the cuts to make larger steaks. To butterfly, cut a double thick steak. Slice the thicker cut into two steaks without cutting all of the way through. Open the two sides to make one larger butterflied cut.

The second way to cut the tenderloin into steaks is the method most used by restaurants. Start by removing the silverskin in the same manner as before. Restaurants remove the secondary muscle called the chain before slicing the tenderloin into steaks.

The chain is the smaller, coarse looking piece that runs the length of the tenderloin. It can be removed by sliding your fingers between the chain and the more solid round filet. You may have to occasionally use a knife but it should come off easily. Add the chain to the thin first cut and the tail piece for use as tenderloin tips or ground tenderloin. Expect a lower steak cut yield with the chain removed.

Beef Butt Tenderloin – Yield 95%

If you are looking for the best value in home-cut tenderloin steak, the butt tender is a better choice. You can remove the silverskin either before or after slicing. The shorter, thicker end is sometimes easier to slice and then trim.

Cost of butt tenderloin is sometimes lower than the PSMO. The demand for restaurant tenderloins pushes up the cost of the PSMOs.

Beef Strip Loin – Yield 77%

Beef strip loin is the easiest of the beef primals to cut at home. The grain of the loin runs lengthwise from one end to the other. Slice the strip across the grain into steaks. If you have a Jaccard®, tenderize the strip across the grain before slicing. Trim the steaks after slicing.

At the top of the steak, on the fat side, you will note a thick piece of gristle about two inches long. Trim this off as well as any extra fat. The steaks will be more flavorful if you leave a fat trim of a quarter inch or less.

Boneless Beef Rib-Eye – Yield 76%

The boneless beef rib-eye is the same cut as the prime rib minus the bone. Many restaurants roast boneless rib-eyes to serve as prime rib because they are easy to slice and serve.

If you choose to make a rib roast out of part of your rib-eye primal, use the fatter "large end". It is called the large end because on the bone, with the top cap on it, it is the larger end of the rib. You can identify it by the kernel of fat in the center of the end. The large end is closer to the chuck and has more internal fat that keeps the meat juicy during roasting.

The rib-eye is easy to cut into steaks. Just slice and trim any excess fat. You can remove the kernel of fat in steaks cut from the large end by poking it out with your finger.

Steaks cut from the leaner end of the rib have the same piece of gristle as the strip steaks. If the gristle is on the outside of the steak, remove it. It is more difficult to remove if it is on the inside of the steak.

Top Sirloin Butt Peeled – Yield 92%

A whole top sirloin primal is a large piece of meat about that weights about 15 pounds. It has a fat covering on one side and has two separate muscles. It has a cap muscle and a larger center muscle. As a whole piece, it is difficult to cut up at home.

The peeled top sirloin butt is just the larger center piece. Most of the fat covering is removed when the cap is cut off. The remaining top sirloin butt is leaner and easier to cut into steaks.

The grain of the peeled top butt runs the short way, across the meat. Slice the top butt across the grain. Because of this, it is more difficult to slice into steaks. It takes awhile to develop the skill level necessary to cut the larger slices. You might find it easier the first time to cut the whole top butt into two pieces before slicing.

You might also notice a thin silverskin-like covering on the inside of the top butt. This is the bone covering left when the top butt is removed from the bone. Trim it off in the same way that you remove the silverskin on the tenderloin.

Beef inside Round Peeled – Yield 93%

The inside round is the primal that is cut into top round steaks and roasts. A peeled top round is one that has a very thin fat covering. Like the unpeeled top sirloin, the fatter unpeeled top round would be a challenge to cut at home. The fat on an unpeeled inside round can be an inch or more thick.

Start by removing the cap muscle to cut the inside round into steaks. The cap starts about three inches from the larger end of the inside round. It is thinner as it starts and gets thicker as it comes off. Using your fingers to pull and separate, use your knife to follow the natural seam to remove the cap. Set the cap aside. The other side of the inside round is covered with silverskin. Slide your knife underneath and remove it in the same manner as removing it on the beef tenderloin.

With the cap removed it is easy to see the grain of the meat. Take

SIMPLE CUTTING TEST

Establishing the true cost, or yield of any subprimal that you cut up at home is very easy. All you need is a scale and the information on the label of the subprimal.

Cost per pound ________

Starting weight ________

Weight of Steaks ________

Weight of Roasts ________

Weight of Ground Meat ________

Other Cuts ________

Other Cuts ________

Total weight of all cuts ________

Divide the total weight of all the cuts by the starting weight. This equals the yield percent.

Divide the cost per pound by the yield percent to determine the actual cost per pound.

The actual cost per pound is for every cut produced. To compare value and savings, compare this cost to the actual cost of these cuts in your local supermarket.

a small slice off the large end to square it up. The first three inches of steak cut from the large end are the more tender cuts. Cut the first steaks thicker and label as grilling steaks when you package them.

The inside round gets smaller across as you slice steaks. Continue to slice steaks until the steaks are a couple of inches across. Trim the steaks after slicing.

The remaining top round and the cap offer choices. You can trim any gristle and process the meat into ground beef. Leave most of the fat on if you are making ground beef. Round is very lean and needs the fat if ground. The trim can also be cut into stew meat cubes or sliced thin for stir-fry.

The top round is also a very good lean roast beef. Leave the fat covering on before cutting into roasts. Cut the inside round with the grain of the beef to make a roast that will be sliced across the grain when served.

The grain of the inside round runs from the thicker end toward the thinner tapered end. Cut the whole inside round into two or three pieces by slicing from the thick end to the thinner end. Trim any visible gristle and use butcher's twine to tie the roasts into a solid shape. These longer roasts can then be cut into smaller roasts.

If you need to feed a crowd the whole top round is an excellent choice. The top round is a lean, easy to slice roast beef. It is a popular choice for community barbecues that grill/roast several large pieces of meat to sell as barbecue beef sandwiches.

BEEF STEAKS

THE right combination of low price and high quality usually trips my trigger to make a steak purchase. I think this comes from my many years of working in a meat market. My family menu each week very closely followed the weekly specials. It was hard not to buy a steak or two if the price was right and I got to hand pick my steaks from hundreds I cut.

Today I eat less steak, but better quality steak. With the added further reprocessing from the beef producers, I now have the choice of many new value- priced quality steaks. Steaks such as flat iron steaks, tri-tips, and chuck-eyes were not widely available ten years ago.

Next to ground beef, steak is probably the most popular beef item on most family's menus. It is quick to fix, and available in a wide range of prices. What beef eater says "no thank you" to steak for dinner?

GRILLING STEAKS

There is a wide variety of steaks that are wonderful when grilled. They range from pricey prime, dry-aged beef steaks to very affordable steaks that can be just as tender.

Tenderness is the first thing most steak eaters use to judge a steak. "How is your steak?" "Oohhh, verrry tender!!" is the response you hope to hear when you serve your guests steak from the grill. After clearing the tenderness hurdle, a great steak has to be juicy and flavorful. The tenderloin or filet mignon is the most tender steak. However, if it is overcooked and poorly seasoned, a simple burger patty tastes better.

Some less tender steaks such as flank steak and skirt steak have such a great natural flavor that their tenderness is not as big an issue. How you slice these steaks when you serve them can affect how tender they are. They both are great choices to add extra flavor with marinades.

How do I choose the best steak? There is a huge variety of steak! Many people already have a favorite cut of steak. Like many meat cutters, a nice rib-eye is my favorite. Ask your local butcher what his favorite steak is. I will bet that he will also say a rib-eye.

For me, eating only one kind of steak is like drinking only kind of wine. I don't think that it would taste nearly as good if a rib-eye was the only steak I ever grilled. I would miss out on steaks that both cost less and have their own terrific flavors and textures.

One way to save money is to buy a thicker cut that can serve three or more people. The serving cost is less than serving each guest a whole, thinner steak. Smaller portions are a more healthy choice to enjoy eating steak more often.

Asking the butcher for a thicker cut than what is on display is a good way to get a better piece of meat. Cutting and trimming one or two nice thick steaks is a different procedure than quickly slicing a whole rib all cut the same thickness. One is production and the other is customer service. Good customer service means giving your customer the best you have available. Many butchers take extra pride in cutting extra thick beautiful steaks.

Some upscale stores display thicker cut steaks in a full service case. Many stores use these cases to showcase their premium cuts. Fancy presentations don't always mean more expensive if you consider the serving cost of a nice thick cut. Consider checking out the gourmet case on your next visit.

Rib-eye Steak

The rib-eye is a popular favorite because it combines tenderness with flavor. As a support muscle it has little connective tissue. Of the steak cuts, it has the most marbling relative to the rest of the beef.

The whole beef rib is cut from the first seven ribs of the front quarter of beef. The large end of the rib is the end toward the chuck. The small end of the rib is the leaner loin end.

In the past, there was a tenderness difference between the large end and the small end of the rib. Most whole beef ribs sold today are just the tender center of the rib with the tougher cap meat removed from the large end.

The boxed beef commodity name for these trimmed center ribs is Golden Lion Ribs or GL Ribs. Rumor has it that they are named after a restaurant called the Golden Lion Restaurant. The specs set for the ribs they purchased are now the standard for most of the whole ribs sold.

A whole GL Rib is pretty uniform in size from the small end (closest to the beef loin) to the large end (closest to the chuck). The small end is leaner and the large end has more internal fat. The additional fat in the large end of the rib adds to the flavor and tenderness of this cut.

The center lean section of the rib-eye is smaller at the end of the large end of the rib. There is a secondary cap muscle that gets larger

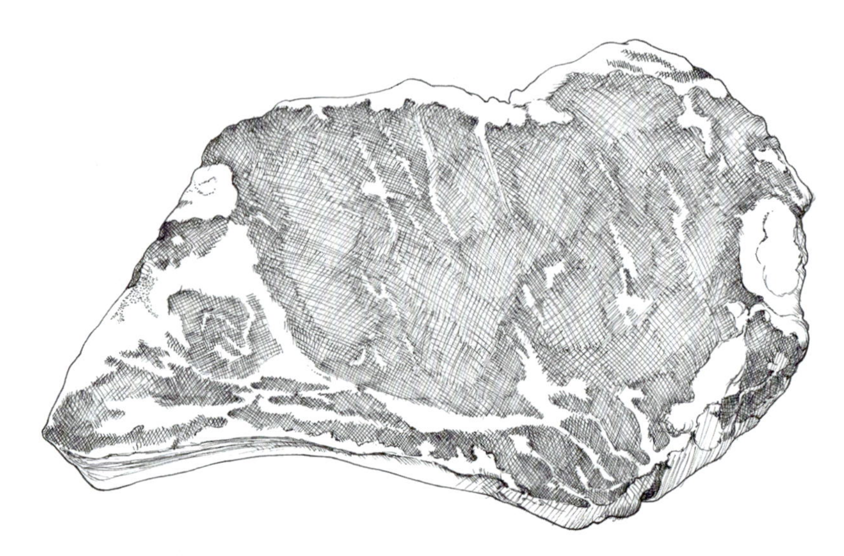

Loin end rib-eye (small end)

moving from the small end to the large end. You can identify where the steak was cut from the rib by how large the cap is. The cuts from the first three ribs have little or no cap meat. Cuts from the last 4 ribs have cap meat ranging from a fourth of an inch to two inches or more.

When buying a rib-eye choose the steak with the most marbling. My personal choice is a cut from the more flavorful large end of the rib. I prefer cuts from the last three to four inches of the large end. These steaks may not look as pretty as the leaner cuts from the small end; however they are the most flavorful. Choose steaks with a top cap of an inch or more.

Cuts from the large end of higher grades of beef can be identified by a small kernel of fat separating the cap from the center eye section. I enjoy eating the cap section of the steak in much the same way as I savor separate bites of tenderloin and strip steak on a porterhouse steak.

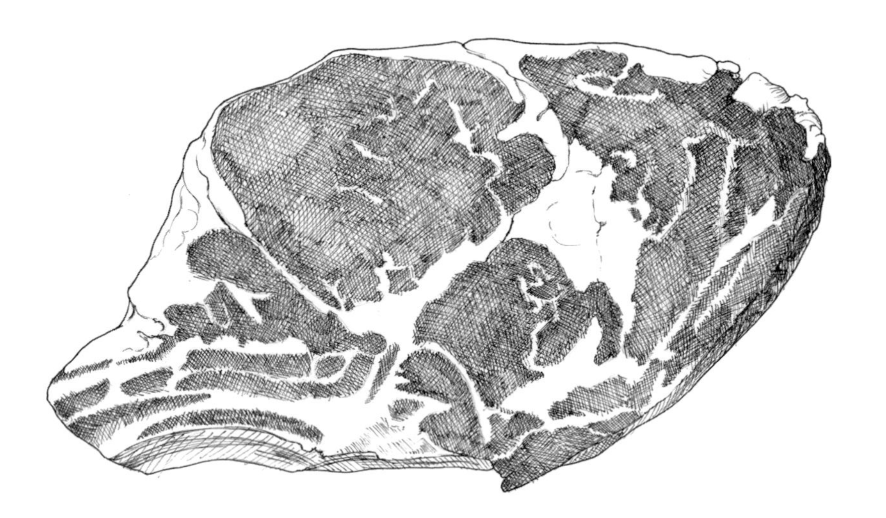

Chuck end rib-eye (large end)

When sold as a steak, the large end of the rib is most often boned out and sold as a boneless rib-eye, or Spencer steak. Removing the bone and the trimming the excess fat helps to make this a more attractive cut. I choose boneless rib-eye steaks in the same manner I choose bone-in steaks. How large is the cap?

Chuck-eye Steak

The chuck-eye is a great value. The rib-eye muscle continues into the first cut of chuck for about three inches. Just a knife blade away from the rib-eye steak, it sells for about 60% of the cost of a boneless rib-eye.

For years, this was the butcher's little secret. We bought them as boneless chuck steak when chuck steak was on sale. If you recognize the shape of the chuck-eye you can buy the first cut, or blade cut of chuck and seam it out.

The first cut boneless chuck steak or roast is usually the smallest, fattest cut in the case. The chuck-eye makes up about a third of the cut. It is nestled in the bottom outside of the cut. Follow the natural seam around the eye to remove it. The remaining meat makes wonderful home ground beef. You can also trim off the fat and cut the lean pieces into a very nice stew meat. Look for chuck blade steaks where the blade bone is white cartilage instead of bone when selecting a bone-in cut. The chuck-eye is in the bottom of the steak, closest to the rib bone.

More often than not today's butcher beats you to it and merchandises the chuck-eye as a value-added cut. Either way, buying a chuck-eye steak is an economical way to enjoy flavor and tenderness similar to a more expensive rib-eye steak. Since there are only two steaks in each chuck, the best time to buy them is when your local retailer advertises boneless chuck steak or roast as a feature. Call your local butcher and order extra chuck-eye steaks next time you see chuck roast or chuck steak on sale.

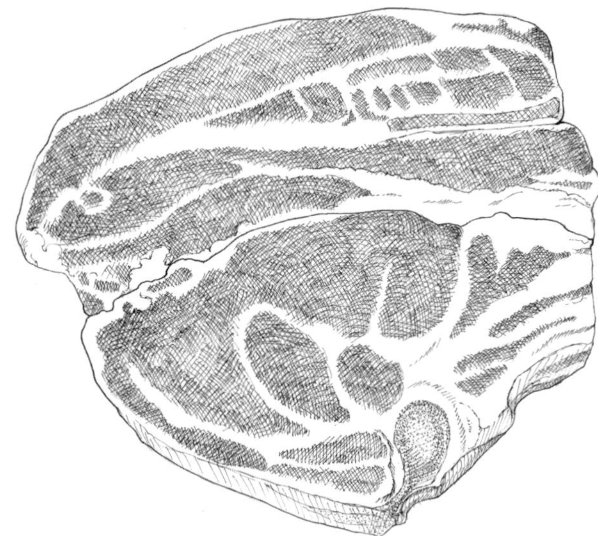

Bone-in blade chuck roast with white cartilage in the blade bone

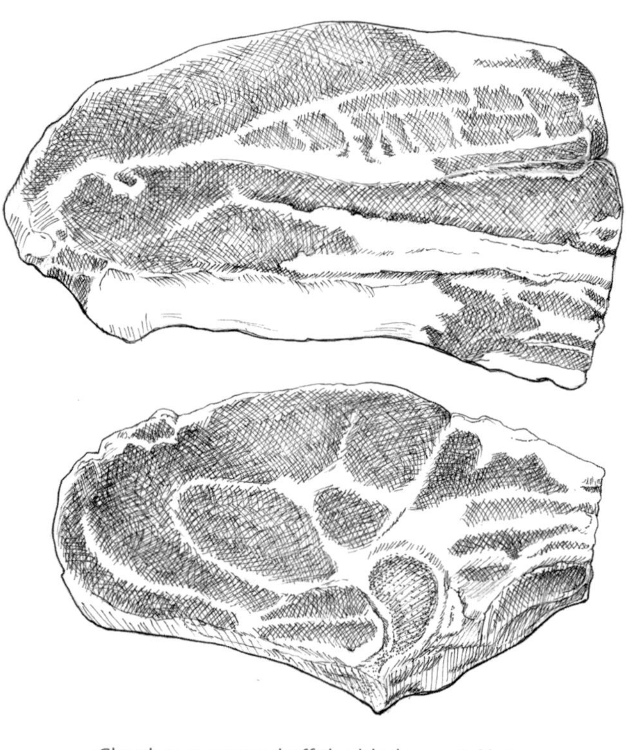

Chuck-eye seamed off the blade roast. Note similarity to a rib-eye steak

Bone removed from chuck-eye, plus lean stew meat. Note: leave some of the fat on if you intend to process the trim into ground beef.

Flat Iron Steak, Top Blade Steak

The flat iron or top blade steak is a relatively new cut of beef. As the beef producers further processed beef to add value, the top blade was recognized as a very tender cut, worthy of its own identity. It probably represents the best value in a tender, lean grilling steak today. Cut from the chuck, it is full of flavor and very tender.

The whole top blade cut is a twelve to fifteen inch long roast, about two by four inches thick, with a piece of sinew running through the middle. With the sinew removed, and the top fat and thin layer of bone fell removed, you are left with two nice flat iron fillets, about one inch thick. The top blade is sold two ways, both as a whole roast and a trimmed flat iron steak.

The flat iron is an excellent choice for a variety of reasons. It is lean, amazingly tender, with a consistent thickness that is easy to grill. Grilled as a whole piece, and sliced across the grain, it is easy to control the serving cost. It accepts all kinds of seasonings and marinades well.

Dollar for dollar, it is the most economical way to enjoy a nice steak. Unfortunately, as it becomes more popular the laws of supply and demand will make it more expensive. I have noticed it showing up on restaurant menus more and more often.

Top Loin Steak, "New York Steak"

The Top Loin Steak is the larger solid muscle of the beef--just a knife cut from the small end of the rib. The beef is separated into the front and hind quarters between the 12th and 13th ribs of the beef. The front quarter is the rib side and the hind quarter is the short loin side.

The short loin is the bone-in primal that is the source of t-bone steaks and porterhouse steaks. The top loin steak is the longer muscle, opposite the tenderloin on a t-bone cut.

Sold as a boneless and a bone-in cut, the top loin steak is a popular grilling choice. Regional names for it include Strip Steak, New York Steak, and Shell Steak. It is leaner than the rib-eye and a very tender cut. The strip steak is a great choice for a flavorful, lean, premium steak. It is not quite as flavorful as a rib-eye because it does not have as much intramuscular fat

Choose a steak with a nice amount of marbling. Leaner steaks can benefit from use of your new Jaccard® tool. Choosing a thicker cut to slice into multiple servings is a good option. The strip steak works well as beef for kabobs.

Tenderloin Steak

The tenderloin is sometimes called the filet mignon. The whole beef tenderloin is called PSMO (Peeled and Side Meat On) tenderloin. The PSMO includes the tenderloin from the short loin as well as the thicker end from the sirloin end.

When the tenderloin is left in the short loin to be cut into t-bones and porterhouse steaks, the thicker end of the tenderloin is still removed from the sirloin end. This thicker, shorter piece is called the butt tenderloin.

Buying tenderloin to grill is not a choice for the budget-minded. It is a nice choice for a special occasion or a special person you would like to impress. There are several classic recipes for beef tenderloin. Sliced very thin, it can be served raw as steak tartar. It is best when grilled from rare to medium in doneness. If it is overcooked it can become very dry.

Tenderloin has a thin shiny skin called the silverskin on the outside of the meat. Removing the silverskin before cooking will make for a more palatable steak. The heat when cooking will cause the silverskin to shrink and the meat to curl. Slide a sharp knife just under the silverskin to remove it.

T-Bone, Porterhouse Steak

If you want to impress your friends or family, serve them each a nice big T-bone or porterhouse steak next time you host the barbecue. They will talk about you fondly for years. If your bank account is not up to buying attention, think of the T-bone as a multiple serving steak. A nice thick T-bone with a large tenderloin section will easily feed two to four people when cut off the bone and sliced before serving.

Choose a thick porterhouse cut with large tenderloin for multiple servings. After the steak rests, remove the tenderloin and the strip from the bone. Slice, and then reassemble the porterhouse on the bone for a spectacular presentation.

T-bone steak is the bone-in combination of the strip steak and the tenderloin steak. The Porterhouse cuts are the first three steaks that have the larger tenderloin side. The tenderloin side of the short loin gets smaller from the porterhouse end to the loin end.

Some stores are unable to get the few extra cents a Porterhouse steak demands. They price all of the cuts from the bone-in short loin as T-bone steaks. Obviously, buying T-bones with a larger tenderloin side is a better value.

Pay attention to how much bone is on the steak when choosing your purchase. The heavier bone will cost you more in loss than a steak with a bit more fat on it.

Most of the bone is at the flat top of the T-bone. This is where the beef is split down the middle into sides of beef. If the split is uneven, it leaves more bone on one side. To enhance their yields, packers choose the short loin side with less bone to process into boneless strips and tenderloins. This leaves the side with more bone to be packed as a bone-in short loin.

Skilled butchers who try to sell the best product will trim the excess bone off before slicing the short loin into steaks. Sometimes if the ad price on T-bones is hot and the butchers are cutting them as fast as they can, trimming the bone does not happen.

Top Sirloin Steak

The top sirloin steak is probably the least favorite steak among most meat cutters. As the steak muscle closest to the back of the beef, it develops more connective tissue than the other middle meats. This makes it is less tender than the rest of the premium steak lineup.

The whole boneless top sirloin has two major muscles. There is a larger center piece and a smaller cap section. The cap portion of the steak has a fat cover along the entire edge of the cap.

The cap muscle when removed is sold as top sirloin culottes. The cap is sliced as a smaller steak or sometimes is cut into stir-fry meat or ready to cook kabobs.

Many stores sell only the center of the top butt with the cap removed. This is another product of the boxed beef commodity program. With the cap removed, the top sirloin is a smaller, leaner, and easier to market steak.

Top sirloin's lower popularity and the basic law of supply and demand sometimes bring the ad price of this steak down to a price that cannot be ignored. Don't be afraid to buy it for the right price.

Top sirloin is usually sold as a boneless cut. Choose steaks at least an inch thick. Thinner steaks can quickly overcook and become tough when grilled. I almost always use my Jaccard® tool when I grill top sirloin steak. Top sirloin is usually very lean with little fat covering. A little extra level of seasoning or a nice marinade adds flavor. Thicker cut top sirloin is a good choice to cut into cubes for kabobs.

Tri-tip Steak

Tri-tip steak is also called bottom sirloin. Don't confuse the word sirloin with the leaner top sirloin. Tri-tip is a much more flavorful cut of meat. It was first popularized by people living in sunny California. They grilled it whole and untrimmed to a delicious moist tender delight. Its popularity was driven by a low, affordable price point. Unfortunately, the laws of supply and demand bring a higher cost today. It is still priced close to the price of top sirloin, making it a very good value.

The whole tri-tip is a thick, triangular piece of meat about two pounds. The grain of the meat runs lengthwise like a flank steak. Slice it thinly across the grain to serve. You can buy it sliced into steaks, but I prefer it grilled as a whole roast.

Look for a roast with small ribbons of fat running between the grain of the meat. If you have a Jaccard® tool, use it on both sides of the meat. Try a whole tri-tip on the grill this summer!

Flank Steak

Flank steaks were an early extra steak offered as the beef producers processed the beef for boxed beef. When stores did their own further processing prior to boxed beef, the flank steak was sometimes added to the ground beef trimmings because there was little or no demand. As customers grew familiar with the flank steak, the demand and the price grew. With only two steaks per whole beef, the demand remains strong. Today, it is among the more popular steaks.

The flank steak is a flat steak about five inches by twelve inches and ¾ to an inch thick. The grain of the meat runs the length of the steak. Slicing it thinly across the grain is one way to make it tender. Asking the butcher to run it through the cuber (a mechanical tenderizer used for making cube steaks) or using your Jaccard® tool also works well.

A very lean steak, the flank steak is a good choice for marinating. It is frequently used in Asian dishes such as teriyaki steak. The Jaccard® tool helps penetrate the meat and makes marinades more effective. A very lean steak, it is best when not cooked past a medium doneness.

Skirt Steak

The skirt steak, sometimes called a diaphragm steak, is a flap of meat along the inside of the bottom of the ribs. It is covered with a membrane on both sides. After the membrane is peeled off, you are left with a long, thin steak fifteen to eighteen inches long, four inches wide, by 3/8 of an inch thick. Like the flank steak, the grain of the beef runs horizontally. Unlike the flank steak, it is very marbled and full of natural flavor. Thinly slicing across the grain helps to increase tenderness.

The most common preparation is Carne Asada, the popular grilled Mexican recipe steak. The steak is marinated in citrus and chili spices and then char-grilled on a very hot grill.

Hanger Steak

If you like steak with a very intense beef flavor, the hanger steak is for you. The hanger steak is not readily available. There is only one on each beef. You may have to search it out in a specialty meat market, but the search is worth it. The hanger steak is sometimes called the butcher's steak. Yesterday's butchers did not offer the steak for sale. They kept it for their personal enjoyment.

The hanger steak is small round piece of meat that hangs off of the short loin, near the heavily fatted beef kidney. It is about eight inches long by two inches in diameter. There is a thin membrane in the middle, running the length of the steak.

Hanger steak can be grilled whole, or split into two pieces, with the membrane removed. It accepts heavy seasoning or marinades well because of its full beef flavor.

A tender steak, it is best when grilled to a rare to medium degree of doneness. I would encourage you to try a hanger steak at least once. You may have a new favorite steak!

Ball-Tip Steak

The ball-tip is another newer cut developed by the beef producers as part of the boxed beef process. Located in the sirloin section of the beef, it is a reasonably tender, lean steak. The whole ball-tip is shaped like half of a whole round ball. It is about eight inches in diameter with a thickness of two to three inches in the center. The butcher slices three to four steaks about an inch thick from a whole ball-tip. As a newer cut it is subject to a variety of names to make it sound more appetizing. Petite Sirloin and Cabrosa Steak are current marketing names.

Use your Jaccard® tool to help make this a more tender steak. It is not bad without being tenderized, however the Jaccard® tool helps. As a leaner steak, it is best if served medium to rare. As a lower priced steak choice, the ball-tip represents a good value. It is an economical choice to serve a larger group some single serving steaks.

Flap Meat Steak

Flap meat is another of the newer steaks marketed by beef producers. It has been available as part of the boxed beef program for quite some time. Supermarkets used to buy it as an easy to use cut for cube steaks and lean stew meat.

In today's meat market, flap meat has been elevated to a steak cut. Cordelico sirloin is one of the names you may see it labeled. The flap meat is similar to skirt steak and flank steak with the grain of the meat running across the steak. It is a small lean cut that comes from the tail of the short loin, or the belly section just behind the ribs.

When I first started cutting meat, we would leave the little tail on the T-bones and fold it around the trimmed steak to add to the weight of the steak. Some butchers just trimmed it off and added it to the ground beef trimmings.

Flap meat steak is a good choice as an alternative to skirt steak. Like skirt steak it can be a little chewy and benefits from marinades. Flap meat is thicker than skirt steak. At the thicker end it can be an inch thick. A whole flap steak will only weigh a pound or so.

Teres Major Steak

Like the flat iron, the teres major is another cut seamed from the chuck. It is a small, lean tender muscle that only weighs 8 to 12 ounces. It is similar in taste but is more tender than the flat iron. Because of its size, it is not as popular as the flat iron cut. The limited availability of product and the name are big obstacles to allowing it to become a mainstream product.

Teres major is not a very sexy food name, so the teres major is usually marketed under another name. Delombre Petite Tender is one of the more popular names. My local chain store sells them individually Cryovac®-packaged in the steak section.

Teres major is a good choice to grill. Simple salt and pepper seasoning is all it really needs. It is a good substitute for almost any premium lean steak cut recipe because of its natural tenderness.

Thin Cut Short Ribs "Korean Style"

I am including a non-steak cut in the list of grilling steaks. Short ribs are a tougher, fatter cut usually braised to tenderness. However, when they are sliced thin across the grain and marinated, they are almost steak-like when served. The tougher chewy texture of the meat is offset by the extra flavor of the short rib. Marinating in a spicy sweet Korean style marinade adds even more flavor.

Sold both bone-in and boneless, Korean style ribs are strips an inch-and-a-half wide by ten inches long and half an inch thick. The ribs have a series of oval bones along one edge. Look for wider ribs when buying bone-in to get a better meat to bone ratio.

Thin cut short ribs can be hard to find in some meat markets. You might have to ask your butcher to cut them for you. Be specific in what you ask for. Ask for cuts from the larger end and ask for them to be cut ½ inch thick. If you just ask for Korean style, you may get any thickness as well as the cuts from the end with more bone than meat.

Beef Ribs

Beef ribs are most often cooked as barbecue beef ribs. A rack of beef ribs is the seven rib section from the whole beef rib. The beef rib is produced when the bone is removed from the rib to make boneless rib-eye roast or steak.

The meat-to-bone ratio of beef ribs is much more bone than meat. Buy at least one pound of beef ribs per serving. Most beef ribs are sold untrimmed. Trim the fat off of the inside of the ribs before cooking. Remove the thin membrane from the inside of the ribs as well. Slide a sharp knife along the bone at one end of the ribs to peel up a section of membrane. You can then use a paper towel to grasp it with your fingers. It should pull off easily. Removing the excess fat and membrane will make cooked product more palatable and easier to eat. Slow cooking will give you a tender finished product.

Stir-fry/Kabob Steaks

Many stores offer precut stir-fry or kabob beef. It offers the consumer a recipe timesaver as well as the opportunity for the store to enhance their profit margin.

Kabobs are sometimes offered for sale with veggies on the kabob with the beef. As a value-added convenience product, kabobs allow the store to sell veggies for meat retails. It makes no sense to pay beef prices for vegetables. Since the beef and the veggies most often cook for different times, it makes sense to grill them separately.

The problem with these cuts is sometimes you do not have any idea of what cut or cuts they used. A better way to go is to choose a cut of meat to use and ask the butcher to cut it for you or to cut it at home yourself. Many butchers will cut a steak into kabobs or stir-fry at no extra charge. Top loin steaks, tenderloin steaks, and top sirloin steaks make great kabobs and stir-fry.

KOREAN-STYLE SHORT RIBS

The sweet, salty, spicy flavor of Korean-style marinade gives these ribs maximum flavor. Marbled cuts such as rib-eye, chuck-eye, or tri-tip are an ok substitution for the short ribs. Just cut them ½ inch thick and follow the same directions for marinating and grilling.

2 ½ pounds of thin cut beef short ribs with the bone in (Serves 4)

Marinade

Olive oil to brush on the ribs prior to grilling

1 bunch of green onions – for garnish, thinly slice the green ends and grill the onions until lightly charred

1 tablespoon of toasted sesame seeds – for garnish.

MARINADE

¾ cup low sodium soy sauce

1/3 cup dark brown sugar

¼ cup rice wine vinegar

1 tablespoon sesame oil

¾ cup of pineapple juice – the pineapple juice helps to tenderize this tougher cut.

2 tablespoons minced fresh ginger Tip: Keep peeled fresh ginger in the freezer and use a cheese grater to mince.

1 teaspoon of crushed red chili flakes – add up to two tablespoons if you like spicy.

3 cloves of minced fresh garlic

Toast sesame seeds in a dry skillet over medium heat for about two minutes. Set aside. Mix all of the marinade ingredients in a small bowl. Reserve ¾ cup of marinade to serve with the cooked ribs. Place the ribs in a zip-lock plastic bag and add the remaining marinade. Squeeze excess air out before sealing. Massage ribs through the bag to coat all of the surfaces with marinade. Refrigerate for three to five hours.

Let the ribs sit at room temperature for 30 minutes prior to grilling. Pat the excess marinade off with a paper towel. Brush the ribs with olive oil. Grill the ribs over an open hot grill for four to five minutes per side or until they have nice charred grill marks. Move the ribs to the indirect heat part of the grill and cook covered for an additional 20 minutes.

While the ribs cook, reduce the saved marinade in a small saucepan over medium heat to about half. Serve the reduced marinade as a side to the cooked ribs.

Garnish with the green onions and the sesame seeds. Serve with sticky rice and a fresh green salad.

BIG CITY STEAKHOUSE STEAK

This recipe is meant to give you that steakhouse experience at home. Choose a well-marbled cut from the best grade of beef you can afford. Dry-aged Prime is the highest grade. Certified Angus Beef® is the highest rated USDA Choice product. Seasoning is important. The addition of compound butter, and/or a house made sauce will complete the steakhouse eating experience.

1 ½ to 2 inch thick steak – any top quality cut will do. (Rib-eye, New York, t-bone, porterhouse, tenderloin, top sirloin) A thick cut steak is important to the finished product.

1 tablespoon of olive oil

2 teaspoons of fresh coarse ground black pepper

1 tablespoons of kosher salt

1 tablespoon of roasted garlic and fresh thyme compound butter

Rub steak all over with a light coating of olive oil. Mix the pepper and salt together in a small bowl. Rub the steak on all sides with the mixture. **Let the steak stand at room temperature for 30 minutes.**

A good sear on the outside of the steak is the first step. If you are using a gas grill, turn the grill up to high and let it heat for at least 20 minutes. Choose mesquite hardwood charcoal if you are using fire. A hot cast iron pan with or without the cooking grid will also work well.

Sear the steak for three to four minutes on each side, turning once. Check the internal temperature with an instant-read thermometer. You want to cook the steak to 120 to 125 degrees for rare and 125 to 130 for medium-rare.

Use indirect heat to finish cooking the steak. Turn off a section of heat on the gas grill, create an indirect section when building the charcoal fire, or finish the steak in a 400 degree oven. The indirect cooking process will be very quick. Check the steak for temperature after three or four minutes.

Remove the steak from the heat and top with compound butter. **Let it rest for five minutes**.

Larger bone-in cuts such as porterhouse or rib-eyes can be removed from the bone and sliced for multiple servings. Re-assemble the steak with the bone and the sliced cuts to serve.

WHISKEY SHALLOT SAUCE

Add a big steakhouse touch to your fancy steak dinner. Start the sauce just after you put the steaks on the grill to sear. The sauce takes about twenty minutes to make. Have your pan hot, your ingredients premeasured, and be ready to go before you start.

2 shallots – thinly sliced

2 tablespoons of olive oil

2 tablespoons of whiskey – I use Irish whiskey

1 ¼ cup of beef stock

¼ cup heavy cream

1 teaspoon of dried thyme or 1 tablespoon of fresh thyme

Kosher salt and coarse ground fresh black pepper

1 tablespoon of cold unsalted butter

1 tablespoon of flour

Sauté shallots and thyme in the olive oil over medium-high heat for two to three minutes or until the shallots start to caramelize. Stir quickly to break up the shallot rings and to avoid burning. Remove from the heat and add the whiskey. Return to the heat and light the whiskey, using either a match or the gas flame.

After the flames die down, add the butter to the pan. Sprinkle the flour in the pan. Cook the flour and butter mixture for a couple of minutes, until it just starts to brown.

Add the beef stock and the cream. Bring to a quick boil, reduce the heat to a simmer, and cook the sauce until it reduces and thickens. Add salt and pepper to taste. Spoon the sauce over the steak to serve. Garnish the steaks with chopped parsley. Recipe makes enough sauce for four steaks.

OPTIONS

Add 3 tablespoons of drained green peppercorns to make a peppercorn sauce.

Add ½ cup of sliced sautéed mushrooms to make a mushroom sauce. Substitute cognac for the whiskey. Add the sautéed mushrooms just before you add the cognac.

COMPOUND BUTTERS FOR STEAK TOPPING

Adding a pat of seasoned butter to any grilled steak before serving is an easy way to upgrade your serving presentation and to add extra flavor. Making a compound butter is a simple process of adding flavor to unsalted butter with fresh herbs and spices. Try making one of the following butters for your next grilling party

ROASTED GARLIC AND FRESH THYME BUTTER

2 sticks of unsalted butter – softened at room temperature.

1 tablespoon of roasted garlic – Cut the top of a whole garlic bunch, place cut side up in a small oven safe dish, add a tablespoon of olive oil to the top of the garlic, roast in a 350 degree oven for 45 minutes. Allow to cool before removing garlic cloves. Mince fine for butter recipe.

1 tablespoon kosher salt

1 tablespoon of fresh thyme – Remove leaves from stems and mince fine.

SMOKY PAPRIKA AND CHILI LIME BUTTER

2 sticks of unsalted butter

1 tablespoon of kosher salt

1 tablespoon of smoky paprika

1 tablespoon of fresh jalapeno pepper – seeds removed and minced fine. (Alternative idea – char the pepper on the grill. Remove the skin and dice before adding to the butter.)

1 tablespoon of fresh lime juice – Grate a teaspoon of zest before juicing. Add the zest to the lime juice.

To make compound butter, blend the softened butter with the seasonings in a small bowl.

Using parchment paper, roll the butter up into a log shape.

Refrigerate the butter for at least two hours before slicing. The butter can be made ahead and will keep for up to a week in the refrigerator. Place one pat on each warm steak for serving. The extra butter can be served with fresh hot bread or rolls.

HOMEMADE TERIYAKI GRILLED FLANK STEAK

Teriyaki marinade and flank steak are a natural flavor combination. Flank steak loves marinades. Making your own teriyaki sauce is easy and adds a personalized touch.

One whole flank steak – Serves four people comfortably. If you have a Jaccard® tenderizer, use it on both sides. Score the steak on both sides with a sharp knife. .

TERIYAKI MARINADE

1 cup of low sodium soy sauce

1/3 cup dark brown sugar

2 cloves of fresh garlic – minced.

1 tablespoon of minced fresh ginger

2 tablespoons of rice wine vinegar

Mix teriyaki ingredients well in a small saucepan. Bring to a simmer over medium heat. Stir frequently until the sugar dissolves (about five minutes). Cool to room temperature.

Reserve one third of a cup for serving over the steak after it is cooked. Place steak in a large zip- lock bag. Pour remaining marinade over steak. Squeeze the air out of the bag before sealing it. Refrigerate steak for six hours to overnight.

Let the steak rest in the marinade at room temperature for 30 minutes. Grill the steak on a hot grill for five minutes and turn it over. Grill on the second side until the internal temperature reaches 125 degrees. Remove from the grill and let the steak rest for six minutes. Slice the steak into ¼ inch or less thick slices. Slice the steak across the grain at an angle so that you get slices that are wider than the thickness of the steak. Top the sliced steak with the reserved marinade and serve.

Steamed rice is a natural side dish for teriyaki flank steak. A simple green salad with sweet rice wine and oil vinaigrette adds a finishing touch.

BALSAMIC HERB MARINATED STEAK

This is a terrific company steak to serve to a small group. Simple to do, it is a steak full of flavor. It looks and smells exotic as it marinates at room temperature prior to grilling. Your guests will be excited to finally get to eat it.

3 strip steaks – Cut one-and-one-half-inches thick and well trimmed. Three steaks will generously serve six adults. Choose one-inch thick steaks for single serving steaks.

Marinade Mixture

2 tablespoons of fresh oregano – Minced fine.

2 tablespoons of fresh thyme leaves – Strip the herb backward to remove just the leaves.

2 tablespoons of coarse crushed black peppercorns – Use the flat bottom of a pan against a hard surface to coarsely crush the peppercorns.

2 tablespoons of kosher salt

2 tablespoons of balsamic vinegar

3 tablespoons of olive oil

Arrange the steaks on a large platter. Whisk all of the marinade ingredients together in a small bowl. Spread mixture on both sides of the steaks and let rest uncovered at room temperature for 45 minutes.

Grill steaks on a hot grill to an internal temperature of 125 degrees for a medium-rare steak. One inch thick steaks will take about five to six minutes per side. Thicker steaks may have to be finished cooking covered, on the indirect heat part of the grill. For a medium done steak, remove the steak from the grill at 135 degrees internal temperature.

Top the steaks with a pat of seasoned compound butter after removing them from the grill. For this recipe I mixed two tablespoons of chopped chives, one clove of finely diced fresh garlic, and salt and pepper with six tablespoons of softened, unsalted butter. As with all compound butters, roll the seasoned mixture into a log on plastic wrap and refrigerate until needed. Compound butters are an easy to make-ahead part of the dish.

Let the steaks rest for six minutes before slicing into ¼ inch thick slices. Spoon any juices over the top.

I like to serve the steak with a fresh green salad and either grilled corn salad or grilled baby red potatoes.

BLACKENED COWBOY RIB STEAK

Searing the bone-in rib-eye steak in an old fashioned cast iron pan is the "cowboy" part of this recipe. I like to use a cast iron grill pan. The raised ridges on the pan leave nice grill marks. Of course you can grill the steak over a nice hot grill as well.

Bone-in rib-eye steak – Ask your butcher to cut between the bones of the rib so that you get a thick steak with the whole bone in the middle. The steaks cut from the large end of the rib are my personal favorite. One steak will serve two or more.

2 teaspoons of kosher salt

1 teaspoon of fresh ground black pepper

2 teaspoons of dark brown sugar

2 teaspoons of smoky paprika – Substitute regular paprika if smoky is unavailable.

1 teaspoon of granulated garlic

½ teaspoon of cayenne pepper

1 tablespoon of olive oil

Using a sharp knife, remove the fat tail of the rib-eye. This process is called "frenching". Follow the natural seam around the bone end of the rib-eye to expose the rib bone. Freeze this fatty morsel to add to lean trim next time you make home ground beef. If you have a customer service butcher, ask him to trim the tail for you.

Mix salt, pepper, sugar, paprika, garlic, and cayenne together in a small bowl. Rub steak on all sides with olive oil. Generously season the steak with the spice mixture. Let the steak rest at room temperature for 30 minutes.

Preheat oven to 350 degrees. Heat a well-oiled cast iron frying pan over medium heat. Sear the steak for four minutes on each side. Place the pan with the steak in the center of the preheated oven. Roast for eight to ten minutes or until the internal temperature reaches 125 degrees.

Remove the steak from the pan to stop cooking. Let the steak rest for 10 minutes. Remove the steak from the bone and slice into ½ inch thick slices. Return the steak to the bone to serve. Pour juices from the pan across the steak prior to serving.

If you choose to grill the steak on a gas or charcoal grill, Use a hot grill with a cooler indirect section to finish cooking after you get a nice sear on the steak.

SKIRT STEAK CARNE ASADA

This is a wonderful way to enjoy skirt steak. Skirt steak accepts marinades well and grills quickly. This is my version of a popular Mexican favorite. Flap meat steak is a good substitute for skirt steak.

1 ½ pounds of skirt steak – This will serve four people generously.

1 tablespoon of kosher salt

1 teaspoon of fresh cracked black pepper

1 teaspoon of cumin – Toast whole seeds until fragrant and grind in a grinder or crush in a mortar and pestle for best flavor

½ teaspoon of dried chipotle pepper (optional) Substitute ¼ teaspoon of cayenne pepper.

4 cloves of fresh garlic – Finely minced.

2 fresh oranges – Juiced plus one tablespoon of orange zest.

1 large fresh lime – Juiced plus one tablespoon of lime zest.

Olive oil – Enough to brush steaks with before grilling.

3 tablespoons of fresh cilantro – Coarsely chopped.

1 jalapeno pepper – Seeds removed and minced fine.

Mix salt, pepper, cumin, and chipotle pepper in a small bowl. Set aside. Cut the steak into serving size pieces.

Juice the oranges and the lime. Add the citrus zest, garlic, chopped cilantro and jalapeno pepper to the juice. Place the skirt steaks in a large zip-lock bag and pour the marinade over them. Push as much air out of the bag as possible and seal the bag. Massage the meat through the bag to distribute the marinade. Refrigerate six hours to overnight.

Let the steak sit in the marinade bag at room temperature for 30 minutes before grilling. Pat the steaks dry and brush with olive oil. Spread the salt and pepper mixture evenly over both sides.

Meanwhile, get your grill as hot as possible. Grill the steak for five to seven minutes on each side, turning once. You want nicely charred grill marks. Remove the steak from the grill and let it rest for five minutes. Slice each piece across the grain, garnish with cilantro sprigs and serve with warm tortillas, chunky guacamole, and sour cream.

CHUNKY GUACAMOLE

Guacamole is a welcome dip or condiment at almost any party. My version is made with chunks of avocado instead of the mashed version.

3 large ripe avocados

1 lemon – Juiced.

2 jalapeno peppers – Seeds removed and diced fine. Leave some seeds in for more heat.

¼ cup fresh cilantro – Chopped fine.

1 bunch of green onions – Sliced thin, including part of the green tops.

¼ cup of olive oil

Kosher salt

Fresh ground black pepper

Using a sharp knife, cut the avocado in half, lengthwise, circling the seed. Twist the two halves to separate. Remove the seed with a spoon. The seed can also be removed by carefully imbedding the knife blade in the seed and twisting it out. If you are comfortable with your knife skills, hold the seed side of the avocado in the palm of your hand and gently but firmly tap the blade of the knife into the seed. It should come out easily with a simple twist of the knife.

Using a table knife, score each side of the avocado into ½ inch pieces. With a spoon, remove the chunks from the skin. Add all of the remaining ingredients, except the salt and pepper, to a large bowl with the avocado chunks.

As you stir the mixture you will notice that it gets creamy as the avocado chunks break down. Continue to stir until the guacamole has a nice creamy texture with noticeable chunks of avocado. Add salt and pepper to taste.

I like this guacamole best when served within an hour of making it. Let it sit at room temperature before serving.

GRILLING AND BRAISING STEAKS

Less tender cuts from the chuck and the round are an economical choice to cook on the grill. Their extra flavor helps to offset their chewy texture. Using a mechanical tenderizer (a Jaccard®) or a marinade with tenderizer (pineapple juice) helps to make them tender.

The slow moist cooking method of braising is a more natural way to cook these cuts tender. Using these cuts to make your own ground beef is also a good way to take advantage of their lower price.

Bone-in Chuck Steak

Bone-in chuck steaks are thinner cuts of bone-in chuck roast. Chuck steak can represent a great value for the summertime griller as the lower demand for chuck lowers the price. Although they can be chewy, chuck steaks have a good amount of internal fat that translates as more flavor.

As a cut, they still contain a couple of tender morsels that are seamed out and sold for more money when the chuck is boned out for boxed beef. The very tender flat iron is one such cut that is sold separately. The first cut blade chuck steak contains the very tender chuck-eye.

The blade bone in the first couple of blade steaks is white cartilage instead of bone. As the steaks are sliced from the chuck, the cartilage turns into a flat bone and then a bone shaped like a backward seven. The steaks cut from this area are called 7-bone chuck steaks. 7-bone steaks sometimes sell for more money because they are leaner and have a better meat-to-bone and fat ratio.

Blade steaks are the best choice to grill. They are the cut closest to the more tender rib steak section. Choose the blade cut where the blade bone is white cartilage. These are the cuts closest to the rib section.

If you have a home meat grinder, buy several blade cuts, seam out the chuck-eye steak to grill separately, and bone out the rest of the steak to make the best ground beef patties you'll have all summer. Unlike ground beef pre-ground elsewhere, I feel comfortable cooking these patties with just a touch of pink in the center when done. In the cooler months the trim around the chuck-eye can be cut into a very lean stew meat.

Boneless Chuck Steak

Boneless chuck steak is more readily available as producers refine boxed beef for added value. Boneless chuck steaks are cut from chuck roll subprimal and chuck shoulder subprimal. The chuck roll is the center of the chuck, minus the flat iron and the outside fat. The chuck shoulder is the lower section of the chuck. It is sometimes merchandised as a cross-rib cut. The chuck shoulder is best when cut into a roast. Although it looks like a nice lean chuck steak, I would not bother buying one to grill.

Boneless chuck steaks cut from the chuck roll are a much better choice to grill. They have a bit more fat but they also have a lot more flavor. The Jaccard® tool helps to tenderize them for grilling. First cut steaks with the chuck-eye still in them are my first choice. However, they are harder to find as more butchers seam out the chuck-eye for extra profit.

Buying boneless chuck steaks to grind your own ground beef is always a good idea. The home ground chuck is much more flavorful than store grind. You have the satisfaction of knowing exactly what is in your ground beef. I like to grill my home ground chuck to a nice juicy medium, with just a touch of pink in the center, something I would never do with store-ground beef. You may never go back to store-ground.

Chuck Mock Tender Steak

The whole mock tender looks very similar in shape to true beef tenderloin. Unlike the tenderloin it is not naturally tender. As a cut from the chuck, it has lots of flavor. It is usually offered as a low priced steak choice. However, there are many better choices for a lower priced steak. The flat iron, chuck-eye, and blade cut chuck steaks are all better choices. For most butchers, the mock tender is a better steak to sell than to buy.

Top Round Steak (London Broil)

The top round steak or inside round steak is the more tender section of the full cut round. It is a very lean steak with a nice beefy flavor. As a grilling steak it needs to be a thicker cut in order to turn out well. Choose a first cut top round at least 1 ½ inches thick. Grilling a thicker steak makes it easier to get a nice sear and to cook it to a nice pink medium doneness.

A first cut top round will be one solid muscle. It is further up the leg end, with less connective tissue than cuts lower on the leg. These first cut top round steaks are more tender than top round cut from the lower part of the leg.

Steaks cut from lower on the leg will have a tougher cap muscle running the length of the steak. The cap muscle will sometimes be a shade darker than the rest of the top round. *When shopping for top round cuts, look for solid muscle cuts, ones without the tougher cap.*

The top round can become very chewy if overcooked with dry heat. It is best when cooked to a medium to medium-rare doneness. The top round accepts marinades or heavy seasonings well. To serve, slice thinly across the grain. The grain of the steak runs vertically. To slice across the grain, make thin slices at a 45 degree angle to the top.

SAUTÉING, PAN-FRYING, AND BRAISING STEAKS

Most of the steaks that are good on the grill are also very good when sautéed or pan-fried. Most often the steaks are cut thinner when they are sautéed or pan-fried. All of the premium steaks such as rib-eye, strip, tenderloin, t-bone, and sirloin are very good when sautéed. Less expensive tender steak choices such as top blade and chuck-eye work well with most cooking applications.

Flavorful, less tender cuts such as bone-in and boneless chuck steaks are wonderful when slow cooked to tenderness by braising. To braise a steak, use a large cast iron pan to brown the steak in oil and then add just enough liquid to keep it moist during the cooking process.

Following are steaks cut from the less tender round that for the most part should not be grilled. They are steaks with too much connective tissue to be tender when cooked with the hot dry heat of grilling. That does not mean they are not good flavored, tender steaks when they are properly cooked. The other benefit is that they represent a good value for your steak dollars.

Full Cut Round Steak

Full cut bone-in and boneless round steaks are sliced from the back leg of the beef. As the slices go further down the leg, they have more

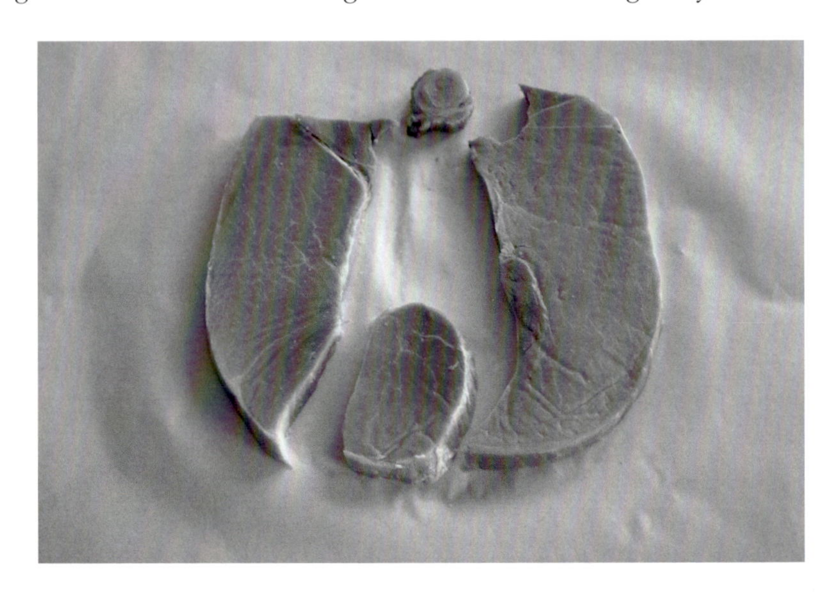

connective tissue as muscles that are used more for locomotion. If you look at the top surface of a full cut round steak you will note that there are natural seams between the three muscles of the cut.

The longer solid muscle is the inside round, or top round. The eye-of-round is the round muscle near the bottom of the steak. The shorter muscle opposite the inside round is the bottom round.

The first cut, more tender steaks have a solid one piece inside round muscle. As the round steaks move down the leg, a secondary muscle, or cap covers the larger inside round muscle. To choose the most tender full cut round steak, choose steaks where the larger inside muscle is one piece. The top round cap is easy to identify because it is a darker shade of red in most cases.

Growing up in my family, the steak my mother usually served was round steak, dredged in flour and pan-fried. I thought it was terrific. Round steak has a very good beef flavor and the addition of a flour coating to keep it moist is still a good idea. If you sauté round steak or if you pan-fry it, use a nice hot pan and don't cook it past a medium doneness. Round steak is best served with just a little pink in the center. The other way to go with round steak is to braise it. Swiss steak is a good example of a way to use braising to make the round steak tender.

Another way to get good value when your retailer offers a sale on full cut round steak is to ask your butcher to cut a five inch thick first cut steak. Asking for the first five inches will get you the best of the whole round. You will get the top of the leg with less connective tissue.

You should get a little or none of the cap muscle in the first five inches. After you get it home you can follow the natural seams to separate the full cut into the top round, the bottom round, and the eye-of-round. These can then be cut into several different products for your freezer.

Slice the top round into 1 ½ inch thick slices for grilling, or slice the top round ½ to ¾ inch slices for sautéing or braising. Slice the eye-of-round into ¼ to ½ inch thick slices for sautéing. The bottom round can be roasted whole or cut into lean stew meat.

Top Round Steak

The inside round or top round is the larger solid muscle of the full cut round steak. It is the more tender part of the whole, full cut round. Choose a top round steak that is one solid muscle. Avoid the top round cuts that have the cap muscle running the length of the steak. The cap is only on cuts from the tougher, lower part of the inside round.

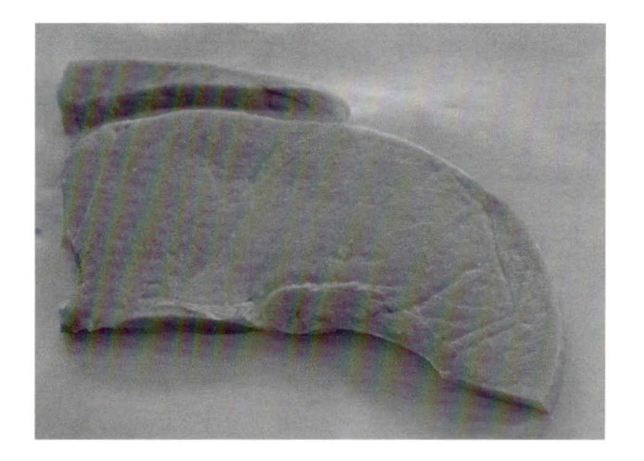

Top Round With Cap

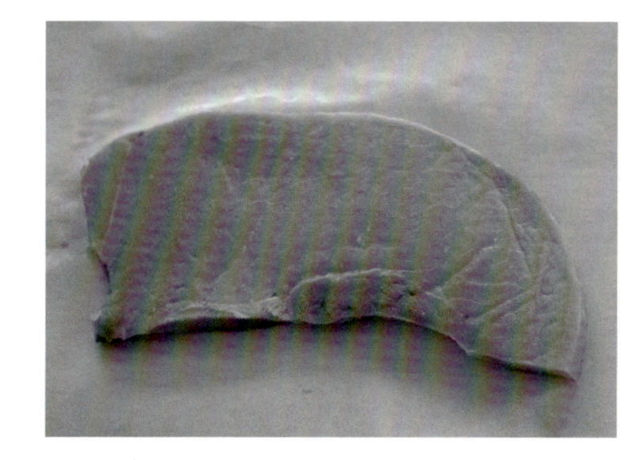

Top Round Without Cap

The top round steak is a nice lean flavorful steak. Thinner cuts can be pan-fried or sautéed. Take care to not overcook. Top round can quickly become dry and tough with over cooking. Top round is a good choice for braised cuts such as Swiss steak. As mentioned earlier, thicker cut top rounds are a reasonable choice for a grilling steak.

Eye-of-Round Steak

The eye-of-round steak is less tender than the top round steak. The nice round shape is probably its best feature. For the butcher, it is a very lean steak, it displays well with a low per serving price. The eye-of-round really makes a much better roast than a steak.

Ask the butcher to "run them through the cuber twice please" to improve their palatability. Like the rest of the round, it can be dry and chewy if overcooked using dry heat. Use a slow moist method of cooking for a tender eye-of-round steak.

Bottom Round Steak

The bottom round is the least desirable part of the round. It makes an OK roast, but is dry, chewy and flavorless as a steak. I honestly do not know any butchers who buy bottom round steaks.

It can be tempting when offered at a low price, but I would go with the chicken instead. Ask the butcher to run bottom round steaks through the cuber or braise them to make them tender.

Sirloin Tip Steaks

If you notice, the full cut round is flat across the top. The other side of the leg is the sirloin tip. The tip is a cut that makes a better roast than a steak of any kind.

The whole tip is most often split into two pieces--the solid side and the loose side. The solid side has little gristle running through it. The loose side has several pieces of gristle running through it.

The leaner, solid side is much like the eye-of-round in leanness and tenderness. The solid side is usually cut into steaks. Some butchers slice the whole tip into steaks. The result is a less than desirable cut. You get the tough lean side along with the even tougher loose side. Butchers who cut these steaks do not buy them.

Boxed beef sirloin tip cuts are removed from the round after the sirloin is removed. Because of this, the whole sirloin tip is cut at an odd angle. Unskilled butchers sometimes slice steaks at the same angle. The result is a steak cut with the grain instead of cutting across the grain. Cutting across the grain shortens the meat fibers to make it tender. Cutting a steak like the sirloin tip in the wrong way just makes a tough cut tougher.

The sirloin tip is a better choice when cut into a roast. It just does not measure up as a steak. It is too lean to be flavorful and too tough to be enjoyed.

Cubed Steak

Cubed steaks are small lean steaks made from lean pieces of steak cuts. As the butcher processes the cuts they set aside these pieces to be made into cube steaks. The other choice for these smaller pieces of meat is to be included in the ground beef trim or cut into premium extra lean stew meat.

Cube steaks have a higher margin than ground beef, so it is a way that a skillful butcher can make more profit dollars. Stores that build a good business by selling well-made cube steaks also buy lean boneless cuts such as flap meat or rib lifters to make additional cube steaks. Well-made cube steaks should be gristle free and not run through the cuber so many times that they have a ground beef texture.

The cuber is a mechanical devise with two rollers much like my grandmother's old washing machine wringer. Each roller is armed with a series of small metal teeth that cut the connective tissue and tenderize the steak as it runs between them. The butcher is able to knit smaller pieces of lean meat into one larger piece by running them together through the cuber.

Cube steaks are a good choice to make a nice "chicken fried steak". Not breaded, they are a quick, easy, small portion steak, great for sandwiches or as a breakfast steak.

Beef Special Trim

The cap meat removed from the beef rib when processing them into GL or Golden Loin style ribs is packaged, along with other lean trim, and sold to retailers as beef special trim. The beef producers remove any visible fat and gristle and then Cryovac® the meat in approximate 15 pound bags as part of the boxed beef program.

You won't find special trim labeled as special trim cuts in the meat case. The retailer uses the special trim to produce a variety of lean cuts. Among the cuts produced from special trim are: beef cubed steaks, extra lean beef stew meat, extra lean beef short ribs, and super-lean ground premium ground beef.

Some wholesale outlets offer special trim for sale to retail consumers. I buy bags of special trim from a local cash and carry store. I use the meat to make my own lean stew meat and extra lean, gristle free ground beef. The cost is usually about half of the cost of store bought lean stew meat and premium ground beef.

If you find yourself in charge of the local community chili feed or have to feed any large group with a slow cooked beef dish, ask your local retailer for a deal on a bag of special trim. The lean cuts are easy to cut into smaller bite-size pieces.

BEEF STEAK STROGANOFF

I have always been the chief cook and my wife Debbie has been the chief bottle washer in our kitchen. It is an arrangement that has worked well for years. I love to cook and she is somewhat of a neat-freak when it comes to cleaning. On the rare times Debbie chooses to cook she is a very good cook. Beef stroganoff is one of her best dishes. I look forward to it every time she makes it (every two to three years). This recipe serves four.

1 to 1 ¼ pounds of thin sliced steak – This dish can be made with any steak from tenderloin to top round. My favorite is a boneless rib-eye. Trim the steak of most of the external fat and slice across the grain into ¼ inch thick slices.

1 ½ cups of sliced mushrooms – Any kind of mushroom will do.

½ of a medium yellow onion – Cut in half and slice lengthwise.

2 fresh garlic cloves – Minced fine.

2 tablespoons of unsalted butter

2-3 tablespoons of olive oil – start with 2 and add more if needed.

1 cup of dry white wine – low sodium beef stock is an OK substitution.

¾ cup of sour cream

Kosher salt and fresh cracked black pepper

Season the steak with salt and pepper.

Heat the oil in a deep sauce pan over medium-high heat. Add the steak and cook for one to two minutes per side, just until it is nicely browned. Remove the steak to a deep plate.

Add the butter, onion, and mushrooms to the pan. Season the mixture with salt and pepper. Sauté for six to eight minutes or until the onions are translucent. Add the garlic and sauté for one minute more.

Add the wine and deglaze the pan for an additional two to three minutes. Add the sour cream. Return the steak and any accumulated juices. Stir and cook for one minute (just enough to heat the meat, tender steak will toughen if it is simmered too long). Taste the sauce and add salt and pepper if necessary. Serve over wide egg noodles or rice.

SWEET PEPPER ROUND STEAK

This is a sweet and spicy version of Swiss steak. The addition of sweet pepper jelly is the kicker. The colorful peppers add eye appeal as well as flavor. The slow braising cooking process tenderizes the steak. The sauce and the tender steak pair well with white rice.

1 first cut top round steak – About 1 ½ pounds. Choose a first cut top round without the secondary cap muscle running the length of the steak. Eye-of-round steak cut ¼ inch thick also works well.

Kosher salt, fresh ground black pepper, granulated garlic

3 tablespoons of olive oil

1 medium yellow onion – cut in half lengthwise and sliced into ¼ in wide slices.

1 red or orange bell pepper – seeded and sliced into ¼ inch wide slices.

2 cups of low sodium beef stock

1 can of diced tomatoes – 15 ounce can. Use the fire roasted ones if you can find them.

¼ cup of red pepper or jalapeno jelly – this is the secret ingredient to the ending flavor of the dish.

Flour – I like to use a cheese shaker jar with large holes on the top. It uses less flour and is less messy.

Cut the top round steak into eight equal pieces. Using a meat pounder, flatten each piece to about ¼ inch thick. Season the steaks on both sides with salt, pepper, and garlic. Dust the steaks with flour.

Heat the oil in a large deep frying pan over medium-high heat. Brown the steaks on both sides and set aside. Take care not to crowd the pan or to move steaks about until they are browned. Brown the steaks in batches to avoid overcrowding the pan.

Add onion and bell pepper to the pan. Add a bit more olive oil if necessary. Sauté onion and pepper for six to seven minutes. Deglaze the pan with the beef stock. Add the tomatoes and the jelly. Stir well. Return the steak to the sauce and cover with sauce. Bring the mixture to a simmer and reduce the heat to low.

Simmer uncovered for one-and-a-half hours. Halfway through the cooking process, rotate the steaks on the bottom of the pan with the top steaks. Keep the steaks submerged in sauce during cooking.

Serve the sauce on the side for spooning over the steak and rice.

MOM'S CHICKEN FRIED STEAK

My mom used to make the best chicken fried steak I have ever had. After years of sampling chicken fried steak at many roadside cafes and diners, I still think Mom's was the best. I have updated her recipe using Japanese panko crumbs. (Something Mom would have used if they were available to her back then.)

1 flat iron steak, approximately 1 ¼ lbs – Ask the butcher to run the steak through the cuber twice. Cut the steak into four equal portions. Pound the steaks to ¼ inch thick.

Salt and pepper

Granulated garlic

Flour – Tip: Keep flour for coating meat in a cheese shaker. The large holes allow you to use just the flour you need without the messy wasting of flour.

1 cup of panko crumbs – You can find them in the Asian section of your store.

2 large eggs

1 tablespoon of cream

1 cup of vegetable oil

Generously season steaks on both sides with salt, pepper, and granulated garlic. Using the cheese shaker, lightly coat both sides of the steak with flour.

Beat the eggs and cream in a shallow bowl, large enough to hold a steak. Spread the panko crumbs on a large plate.

Dip the steaks in the egg mixture. Coat both sides with the panko crumbs. Place the breaded steaks on a cookie sheet lined with parchment paper and refrigerate for at least 30 minutes. This will help the coating stay on during cooking.

Heat the oil in a deep frying pan over medium to medium-high heat. (I prefer a large cast iron frying pan.) When the oil reaches 375 degrees you are ready to cook. The oil temperature will lower to 350 or so as you start to cook the steaks.

Fry the steaks in batches for four to five minutes per side or until golden brown. Lower the steaks in the hot oil away from you to minimize splatter. Keep the cooked steaks on a rack placed over a baking sheet in a 200 degree oven until served.

Make the following sausage gravy to serve with the steak. I prefer to serve the steak with the gravy under the steak, rather than the gravy poured over the steak.

SAUSAGE AND MILK GRAVY

½ pound of pork breakfast sausage

½ small yellow onion cut into medium dice

1 to 2 tablespoons of unsalted butter

2 tablespoons of flour

2 cups of whole milk

Kosher salt and fresh ground black pepper to season

Using the same pan that the chicken fried steak was cooked in, sauté the sausage and onions until the sausage is browned. Pour off the oil used for frying the steak and brush out any leftover bits of coating with a paper towel before starting to cook the sausage. Remove the cooked sausage and onion mixture to a small plate.

Add as much butter as necessary to have at least two tablespoons of fat. You need equal parts of fat and flour to create the rue. Rue is the thickening agent of the gravy.

Stir the flour into the fat until it is smooth.

Turn the heat to medium-high and add the milk. Bring to a boil, reduce the heat to a simmer and stir until the sauce thickens. Add the sausage and onions to the gravy.

Generously season the gravy with fresh ground black pepper. Taste the gravy and add salt if needed.

Serve the gravy with the chicken fried steak.

You can substitute turkey or chicken sausage for the pork sausage. The gravy is also very good when served as biscuits and gravy. Top the biscuits and gravy with a fried egg for a hearty, old-fashioned breakfast.

To make simple white gravy, leave out the sausage and add an extra tablespoon of butter and flour.

BUYING BEEF ROASTS

FOR many people buying a beef roast is an even more intimidating decision than choosing a steak for dinner. Most meat customers have more experience cooking steak or ground beef than cooking a roast. Unfortunately, many have also had an unsuccessful roast cooking experience. Knowing how to choose and cook a roast beef greatly enhances the meat eating experience.

Roast cuts are thicker or whole cuts of the same beef that is sliced into steaks. Just like steaks, premium cuts are best cooked with dry heat, and cuts with more connective tissue are made tender with moisture. A difference is that many lean cuts from the round are best when dry roasted to a juicy medium doneness instead of roasted with moisture.

All of the premium steak cuts make terrific roasts when cut in a larger roast form. The taste and texture of these cuts when roasted is much different than when they are cooked as steaks. The golden lion rib that the rib steaks are cut from becomes prime rib when roasted. The prime rib has long been thought of as "The King of Roasts". The beef tenderloin becomes chateaubriand when roasted or beef wellington when covered with pâté and a flaky pie crust.

When a customer would ask me which roast do you recommend? Just like a good car salesman, I would start with the most expensive and work down from there. "A prime rib of course" would be my answer. Seriously, for most of us buying a prime rib is a special occasion purchase. It is a wonderful centerpiece for a holiday or celebration dinner. It is loaded with fat and calories so it is probably not good for our long-term health to eat it on a daily basis. Oh well, what a way to go!

Rib Roast

The whole beef rib is cut from the first seven ribs of the front quarter of beef. Virtually all of the boxed beef ribs sold today are golden lion style ribs. The cap- off golden lion or GL rib is much easier to process into steaks or roasts. Retailers and restaurants buy the GL rib in bone-in and boneless versions.

Before the golden lion style rib became the standard rib sold, beef ribs were marketed with the tougher cap meat and fat covering intact. The whole rib was much larger at the end with the cap. The rib was sold as small end or large end roasts. The small end (the loin end of the rib), was the more expensive, preferred cut. The large end (the chuck end of the rib), was the lesser cut and still had the tougher cap meat.

The terms "large end" and "small end" are still sometimes used by old school butchers to identify the cut of prime rib roast. The reality is that with the golden lion rib, the fatter chuck end of the rib is often a smaller cut than the loin end.

Asking for the large end or the small end cut when shopping for a rib roast will only confuse most of today's less-skilled butchers. Ask for a loin end cut if you prefer a leaner roast. Ask for a cut from the fatter chuck end if extra flavor is your preference.

To decide how large of a rib roast you need to buy start by counting how many people you need to serve. Do you want left-overs? Nothing is better than the French dip sandwiches enjoyed the day after your guests have gone home. A whole seven-bone beef rib is about 15 pounds. As a bone-in cut it will serve fifteen people generously. To serve fewer people, order the roast by the bone. One rib will serve two to three people.

The term Prime Rib does not refer to the grade of the beef. It is a restaurant term for the cooked cut. Look for the grade of beef on the label. Remember, that grading of beef is optional. Only USDA graded beef is allowed to include the grade in the description.

The grade of beef makes a great deal of difference in the quality of the finished cooked rib roast. The internal fat marbling in the better grades of beef add flavor and tenderness that make a great prime rib memorable. Buy the highest grade of beef your budget will allow when purchasing a rib roast.

The best time to buy a rib roast is during the Christmas and New Year's holiday season. Many stores lower their ad prices to rock bottom during this competitive time. Some stores will also sell you a rib roast at ad steak prices during the summertime grilling season.

I would bet that most of the prime rib consumed across America is eaten in restaurants. From steakhouses, to casino buffets, to the neighborhood diner, we all have our favorite place to eat prime rib. Many of these prime rib eaters have never roasted their own prime rib. Investing a lot of money in a cut you fear you might ruin by cooking wrong is enough to scare many people away from buying a rib roast. The reality is the rib roast is one of the easiest roasts to cook.

There are a variety of ways to cook a rib roast. You can roast it slowly at a low temperature, turning up the heat at the end to get a nice crust. You can start it at a higher temperature to sear, and finish at a lower temperature. You can encrust it in rock salt and roast slowly. All of these methods work well.

The only real key to success is to use a good thermometer and to pull the roast from the oven eight degrees earlier than your desired finished temperature. Tent the finished roast with aluminum foil and let it rest for 20 to 30 minutes.

Easy to serve Holiday Rib Roast

Cutting the bones off the rib and trimming the interior kernel fat helps to make your finished prime rib easy to serve. After the roast is done, simply remove the strings and remove the roast

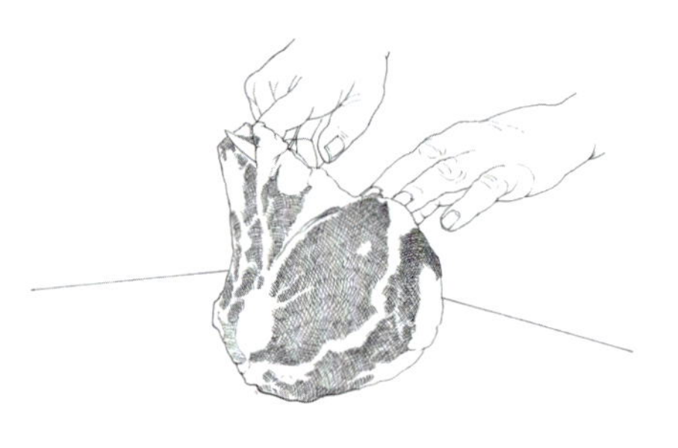

Step 1

Step 2

from the bones. You are left with an easy to slice boneless roast. If you do your own, an added benefit if that you can add seasoning or extra garlic cloves to the center of the roast before you tie it back on the bones.

Step one: Turn the roast with the ends of the rib bones up. Cutting with a sharp knife, slide along the bones to separate the roast from the bones.

Step two: Place the rib with the fresh cut surface up. Following the natural seam, use your knife to peel back the top cap, exposing the kernel of fat in the center of the roast. Again, following the natural seam, remove the fat from the center. **Note; Step two is for the large end only. If you are cooking a small end roast, proceed to step three.**

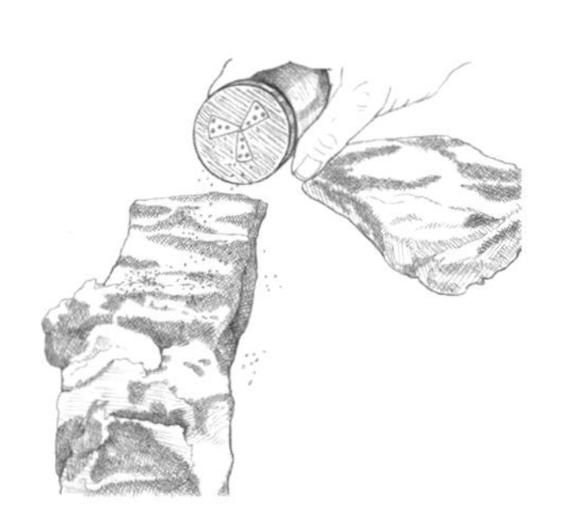

Step 4

Step 5

Step three: Trim the excess fat from the tail of the rib. Be careful to not over trim. You need at least ¼ inch of fat to help in the roasting process.

Step four: This is the time to add seasonings to the inside of the large end if you choose. A nice coating of olive oil will help to make your dry seasoning stick.

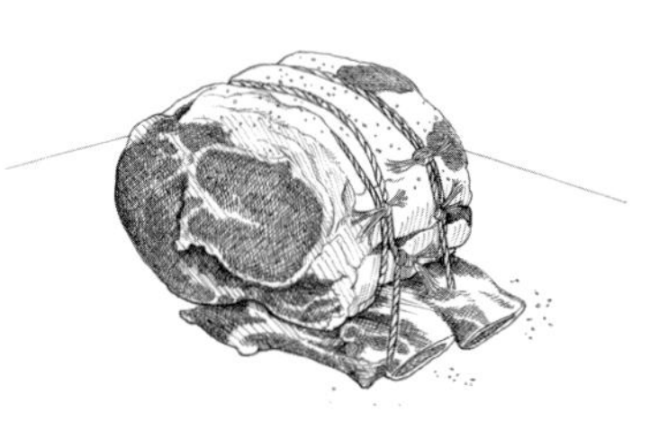

Step 6

Step five: Roll the trimmed roast back together. Push the open center (where the fat was removed) back together. You don't want a hole in the center of the roast. Using butcher's twine, tie the roast. Generously season the outside of the roast.

Step six: Place the trimmed boneless roast back on the bones. Using butcher's twine, tie the roast back on to the rib bones. **Note:** Season the rib bones on both sides before you tie the roast back on. Serve the bones on the side or save them for holiday football snacks.

To tie the roast, cut several strings about eight inches longer than needed to go around the roast and bones. (Space the strings a couple of inches apart). Start by making a simple over and under knot. (Just like the knot when you start to tie your shoelaces). Wrap one end of the string under one more time, and pull both ends tight. Hold the knot tight by looping both ends over and under in the same way you would finish a conventional knot.

The rib roast is ready to roast. Roast in the same manner as an untrimmed rib roast. The trimmed roast is easier to serve when done.

EASY, MEMORABLE PRIME RIB ROAST

After trying many different ways over the years, I have used the following method for the last fifteen years. Although you can roast a boneless prime rib, I prefer the bone-in cut. The quality of the beef is the biggest determining factor in memorable prime rib. A well-marbled cut will be noticeably more tender and flavorful.

Bone-in beef rib roast – Ask your butcher to remove the bones from the roast, remove the kernel fat from the large end, and tie the trimmed roast back on the bone. To do this yourself, follow the steps in the Easy to serve holiday roast on pages 68 and 69.

Olive oil – enough to coat entire surface of roast

Kosher salt, fresh cracked pepper, granulated garlic

SEVEN EASY STEPS TO MEMORABLE PRIME RIB

Step one: Rub olive oil over all surfaces of the roast. Season the roast generously with salt, pepper, and garlic. It is OK to use more than you think you need because the seasonings are just a crust on the outside of the roast.

Step two: Place seasoned roast bone-side down on a roasting pan. Leave at room temperature for one hour. This is key to a larger pink center in the finished roast.

Step three: Insert an oven-safe thermometer into the center of the roast. Place it where it is easy to read without removing the roast from the oven. Be careful to not touch the bone with the thermometer.

Step four: Place roast in the center of a preheated 400 degree oven and roast for 20 minutes.

Step five: Turn oven down to 350 degrees. Do not open the oven.

Step six: Using 16 to 18 minutes per pound, check the roast for internal temperature. Remove the roast at 125 degrees for a nice medium-rare. Pull the roast at 130 degrees for a medium doneness.

Step seven: Allow the roast to rest for 20 to 30 minutes before serving. The roast will continue to cook as it rests, gaining an additional 8 to 10 degrees. Resist the temptation to slice a taste. Any cut gives juices a place to escape.

Combine ¾ cup of sour cream with ¼ cup of creamed horseradish and the juice of ½ a fresh lemon to make a welcome condiment to the roast.

Premium Steak Cuts as Roasts

In addition to the beef rib, the **top loin, tenderloin, and top sirloin** all make great special occasion roasts. Dry roast them, fat-side up. Overcooking is probably the worst thing you can do to them. They become dry and flavorless when overcooked. Medium to medium-rare, they are juicy and flavorful. If you like your roast well done, save your money and choose a less expensive cut.

Roast premium steak cuts at a higher temperature of 375 degrees or more. Because they are already tender, they don't really benefit from low and slow roasting.

Pan-sear smaller cuts in a hot, oiled pan and create a nice crust prior to roasting. The roasting time is very short (as little as five or six minutes) for smaller premium cut roasts.

The **top loin** is rarely thought of as a roast cut. Since this is a very popular steak larger roast cuts seldom show up in the meat case. The big box discount stores sometimes sell whole strips in the Cryovac® section of the case. Even these are most often cut into steaks after purchase.

My favorite way to roast the top loin is to split the top loin roast down the middle before roasting. This gives you two long roasts similar in shape to the beef tenderloin. After seasoning, sear each piece on all sides before finishing the roast in a hot oven. It will roast very quickly. Check the internal temperature after five to six minutes in a 400 degree oven. Remove at 125 degrees for a nice medium-rare. Slice into ¼ inch thick slices after the roast rests. Serve it at room temperature as the beef entrée in a party buffet.

The **beef tenderloin** is a very special cut when roasted. It's nice round shape and melt in your mouth tenderness make a very elegant presentation when roasted to a medium to medium-rare doneness.

Remove the silver skin from the tenderloin before roasting. The silverskin is a tough silver membrane that will shrink as the tenderloin roasts, causing it to curl.

To remove the silverskin, slide a sharp knife just under it with the blade turned slightly up. Run the blade from one end of the tenderloin to the other end. You may have to do this more than once to remove it all.

Tie the tenderloin, with string every couple of inches to keep it nice and round. The tenderloin is a good cut to pan roast. Season the roast and brown on all sides in a heavy skillet. Finish cooking the tenderloin roast in the oven. Preheat the oven to 425 degrees. It will cook quickly, in as little as five minutes. For a medium-rare roast, pull the roast from the oven just as soon as the internal temperature reaches 125 degrees.

The **top sirloin** makes a nice roast. Like the other premium steak cuts, it is not marketed frequently as a roast. Many stores sell only the center of the top butt with the cap removed. This center muscle is very lean, with little fat covering. If you want to try a top sirloin roast, ask the butcher to tie it in a uniform shape with some additional fat covering.

Roast the top sirloin roast fat-side up in a 375 degree oven. It is not necessary to pan-sear the top sirloin roast prior to roasting. The roast is large enough to brown during the cooking process. The top sirloin has more connective tissue than the top loin and tenderloin. It can be a bit chewy if cooked to a rare doneness. Allow the internal temperature to reach 130 degrees before pulling the roast from the oven.

The only time that I would consider buying a top sirloin roast is when the store has a hot price

> If you like your premium cut roast cooked well done, save your money and choose a less expensive cut.

on USDA Choice or higher beef. A leaner, lower grade of beef with little marbling will be no more tender than a well-marbled lean top round roast.

BEEF ROASTS CUT FROM THE ROUND

The round is the back leg of the beef. As a locomotion muscle, it has more connective tissue than the top of the beef where the premium cuts are located. You might assume that you need to roast all round cuts with slow moist heat. As steak cuts, slow moist heat is helpful in making round cuts tender. However, many round cuts benefit from a dry roasting process. Because they are very lean, roasting them to a juicy medium-rare doneness is sometimes the best choice.

For every bottom round produced there is a top round. I would pass on the bottom round sale and wait for the top round to go on sale.

Top Round Roast

The top round or inside round is the best part of the beef round. As a roast, it is a lean, easy to slice candidate for dry roasting. It is a good value choice for a wonderful weekend dinner with leftovers for French dip or Philly Cheese Steak sandwiches later in the week.

The best cut of the top round is the first cut of the top round. This cut is sometimes merchandised as a watermelon cut top round roast because of its shape. The first cut has a shape similar to a wedge of watermelon. Take note of how the grain of the meat runs so that you slice across the grain to serve.

Choosing the best part of the top round for a roast is the same as choosing the most tender top round steak. Look for cuts without the tougher cap as part of the roast.

Because of how the grain of the beef runs, skilled butchers slice the top round into roasts by cutting it with the grain. This way, when you slice the roast after roasting, you cut across the grain for a more tender serving. The front part of the whole top round is one solid piece and slices nicely. Roasts cut from the other end have less fat covering and have more of the cap muscle. Don't let more fat covering discourage you from buying a better roast. You need the fat covering to keep the roast moist during roasting.

Bottom Round Roast

The bottom round is the least tender part of the round. The exception is the thicker pointed end of the bottom round. It is usually separated and sold as a boneless rump roast. The rump roast can be either dry roasted or slow cooked as a braised roast. The rest of the bottom round roast needs to be slow cooked as a braised roast to make it tender.

Given that for every bottom round produced there is a top round, I would pass on the bottom round sale and wait for the top round to go on sale.

Eye-of-Round Roast

Even most butchers under-appreciate the eye-of-round roast. It took me several years in the meat business to finally buy one. After the first one I was sold on its value.

At first glance it looks like a tiny roast hardly worth cooking. Its size and shape is actually its best feature. It roasts quickly, and its solid round shape slices nicely for a variety of roast beef dishes. It is a great choice for small families and singles to enjoy fresh roast beef.

Choose a roast with a nice fat covering and roast uncovered at 400 degrees to a finished internal temperature of 130 degrees. A two pound roast will take less than one hour to cook. Slice it thinly across the grain to maximize its tenderness.

Leftover eye-of-round roast is easy to slice for hot and cold sandwiches. It is better than the precooked roast beef sold in most delis.

Sirloin Tip Roast

If you have ever purchased a full cut round steak, you might notice that the top of the steak is flat. What is missing from the other side is the sirloin tip. The boxed beef term for sirloin tip is beef knuckle.

The whole sirloin tip weighs eight to ten pounds. The butcher usually splits the whole sirloin tip lengthwise down the middle to process into roasts. One side is the solid side or silver side (because of a thin covering of a silver membrane.) The solid side is one solid muscle. The solid side is the best choice as a roast. You can identify it by its solid appearance at the end of the roast.

The other side is called the loose side. The loose side is the less desirable of the two roasts. It has a thick piece of sinew and more gristle running through it. After trimming, the roast has a separate flap of meat and is tied into one roast.

As an ad feature, the sirloin tip roast represents a good value. It is a nice lean roast. Dry roast it to a juicy pink medium doneness (take it out of the oven when the internal temperature reaches 135 degrees). Slice it thin across the grain for maximum tenderness.

Seasoned, ready to roast eye-of-round…recipe on page 76

DELI-STYLE EYE-OF-ROUND ROAST

You cannot buy deli roast beef this good (unless you pay an arm and a leg for it). The eye-of-round is a perfect choice for easy to slice cold, sliced roast beef. Choose a larger cut for a nice Sunday roast beef dinner, and follow with several easy weekday dinners.

3 to 4 pound eye-of-round roast – tell the butcher that you want the thicker end. The thinner end won't roast evenly. Ask for it untrimmed. You want the fat covering left on.

3 tablespoons of whole peppercorns

3 tablespoons of kosher salt

1 tablespoon of granulated garlic

3 tablespoons of fresh thyme

2 tablespoons of whole mustard seed

Olive oil

With a sharp knife, score the roast in a ¾ inch grid, just through the fat covering. Let the roast rest at room temperature for at least an hour before roasting.

Preheat the oven to 400 degrees.

Crush the pepper corns. (I use a mortar and pestle) You can use a heavy pan and any hard surface. If you are using a mortar and pestle, add the fresh thyme to the pepper after it is mostly crushed. If not, mince the thyme with a knife.

Add the salt, garlic, and mustard seed to the pepper corn and thyme mixture. Mix well.

Spread the mixture evenly on a baking sheet. Lightly coat the eye-of-round roast with olive oil. With the fat-side down, roll the roast in the spice mixture.

Roast the eye-of-round fat side up (uncovered) to an internal temperature 120 degrees. At this high temperature the roast will cook quickly--in one hour (+ or – ten minutes). Because of the small diameter of the roast and the high cooking temperature, the roast will gain 10 to 15 degrees after you take it out of the oven.

Allow the roast to rest for 20 minutes or until it is cold before slicing. Slice the roast as thin as possible. Expect four servings per pound for the best hot and cold beef sandwiches you have ever had. French dip sandwiches, hot roast beef sandwiches, cold roast beef sandwiches, barbecue beef, any way you would use deli roast beef.

BEEF ROASTS CUT FROM THE CHUCK

The chuck is included in the front quarter of the beef. Cuts from the chuck have more internal fat than cuts from the hind quarter. As we know, fat equals flavor. Most chuck cuts are best when braised and slow roasted to tenderness.

Bone-in Chuck Roast

The boxed beef standard for bone-in chucks is a neck off, square cut chuck. This means that it is ready to slice from end-to-end into bone-in chuck roasts. The butcher starts at the blade end (the end closest to the rib roast) and slices the chuck into roasts.

The first cut blade roast is just a knife blade away from the large end of the rib roast. You can identify the first cut blade roast by looking at the blade bone. The long flat bone will be cartilage instead of bone. The center of the first cut blade roast contains the tender chuck-eye. It can be seamed out and cooked as a very tender steak cut. The rest of the blade roast can be boned out and either cut into stew meat or made into ground beef.

If you are looking for a bone-in chuck roast for roasting, the seven-bone chuck roast is a better value. It has more meat-to-bone and less fat than the blade roast. Braised and slow roasted, there is little difference between the tenderness and flavor of the blade and seven-bone roast. The seven-bone roast is identified by seven shaped bone, not the number of bones in the roast.

Boneless Chuck Roast

The boxed beef standard for boneless chucks is the neck off chuck roll. The chuck roll is just the very center of the chuck. The beef packer bones the chuck and merchandises the chuck mock tender, teres major and flat iron as separate cuts.

With the increasing demand for tender flat iron cuts, the boneless chuck is becoming the norm in a chuck roast. It is getting harder to find a bone-in chuck roast in some stores.

The whole chuck roll is easy for today's butcher to process into roasts or steaks. For the most part it is simply sliced from end to end. The first cut blade end of the chuck roll contains the tender chuck-eye. Just like the bone-in blade roast, it can be boned out and cooked as a very tender steak. If your butcher displays a section of chuck-eye steaks during a boneless chuck sale, chances are he has beaten you to the punch. Buy them anyway. They are still a good value.

Shopping for boneless chuck roasts is easy. The leanest roast in the size that fits your needs is the best choice. Cooking them is even easier. A slow braised roast fills your home with the smells of dinner like Grandma used to make. Boneless chuck is also a great candidate for slow roasting and shredding for barbecue beef or Mexican dishes.

Round Bone Chuck Roast

The round bone roast used to be a very popular roast when butchers processed whole front quarters in the store. As part of the change to boxed beef the round bone roast is now processed into a boneless shoulder clod roast. The round bone roast was a premium chuck roast because of its fat-to-bone ratio. It is a very lean roast with one small round bone in the middle. If you see one for sale at your local, small, old fashioned butcher shop, give it a try. Braise it as you would any good chuck roast.

Chuck Cross Rib Roast

The cross rib roast is another roast lost to the process of boxed beef. It is cut from the end of the round bone roast. The cross rib is a round, log shaped, boneless roast formed from a square bone-in cut.

Like the round bone roast, the cross rib roast is not usually available in most retail meat markets.

Chuck Shoulder Clod Roast

The boneless shoulder clod roast is a combination of the same meat that used to be processed into round bone roasts and cross rib roasts. Some butchers use string and tie the shoulder clod into a rolled boneless chuck roast. Others slice the thicker end of the clod into flat cut roasts.

Sometimes the smaller end is rolled and labeled as a cross rib roast. This version of a cross rib roast is missing the flavorful boneless short rib piece that was tied on the pre-boxed beef version. In boxed beef, the short rib piece is removed and sold as bone-in short rib.

However it is cut, the shoulder clod is a good choice as a lean, boneless chuck roast. It has much less internal fat than the chuck roll. Choosing a shoulder clod is more about choosing the right size than the right cut from the clod.

The shoulder clod is best when braised like any other chuck roast. You can dry roast a rolled roast to a medium doneness, however I think a roast from the round is a better choice if you want a lean, medium-rare roast.

Brisket

The beef brisket long ago separated itself as a special cut of beef. It is the first choice for corned beef, for pastrami, and for many beef barbecue aficionados. All of the methods of cooking brisket use a moist, low and slow cooking process.

The fresh brisket is sold as a whole brisket, both trimmed and untrimmed. It is also cut in half and marketed as a flat cut and the point cut. The flat cut is the leaner choice. It is one solid piece of lean meat with a fat covering on the top. The point cut has a wedge of fat separating two solid pieces of meat. If you are choosing between the two, the flat cut is the better value.

The grain of the beef is easy to recognize on a beef brisket. It runs lengthwise, not unlike a flank steak. It is important to slice the brisket across the grain for maximum tenderness. The cooked fat on a flat cut brisket is easily removed before slicing.

Many of us enjoy brisket as corned beef or its smoked cousin, pastrami. Who doesn't like a deli sandwich piled high with corned beef or pastrami? The folks in Texas have made barbecued brisket a national favorite. This built-in demand makes brisket a rare ad special in most meat markets.

St. Patrick's Day corned beef sales are the consumer's best time to save money on beef brisket. It is usually packaged in a vacuum-pack with a fairly long shelf life. Check the "sell-by" date on the package and buy extra.

If fresh brisket is not on your regular shopping list, give it a try. There are tons of recipes for braised and barbecued brisket. Even at regular price it is an economical choice to feed a crowd at your next barbecue.

Beef Top Blade Roast

The beef top blade roast is the same tender cut as the flat iron steak. The difference is that when it is sold a roast, it still has the thick piece of gristle running through the middle of it. Seam the gristle out of the center and you will be left with two flat iron steaks.

Slow cooking in a nice braise will dissolve most of the center gristle. However, my favorite way to enjoy the top blade as a roast is to remove the gristle without cutting through the two flat irons, and to roll it up like a jelly roll stuffed with a nice mushroom stuffing. Perfect for special occasions or holidays, you will be amazed at what a flavorful and tender roast it is.

MUSHROOM STUFFED FLAT IRON ROAST

This tender cut, normally served as a steak also makes a wonderful roast. As a company dinner roast, your family and friends will feel they are being treated to a special dinner. The recipe is for one flat iron. However, instead of cutting one piece into two you can use two whole steaks, and double the stuffing, to serve six to eight people.

1 flat iron steak – a 1 ½ pound cut will serve four.

½ pound of fresh mushrooms – Sliced thin. You can use whatever type of mushroom you like best.

2 large shallots – Sliced thin, you can substitute 3 tablespoons of minced red onion.

1 tablespoon of fresh thyme leaves – Coarsely chopped.

1/3 cup of shredded parmesan cheese

2 tablespoons of unsalted butter

3 tablespoons of olive oil – Two for sautéing the mushrooms and shallots, and one for rubbing on the outside of the roast before seasoning.

Kosher salt, fresh ground black pepper

Leave the roast out at room temperature while starting the stuffing mixture. Preheat the oven to 350 degrees.

In a saucepan over medium heat, melt the butter with the olive oil. Add the sliced mushrooms and shallots and sauté for five to six minutes or until the mushrooms are cooked. Season the mushrooms with salt and pepper as they cook. Add the thyme and the garlic, cook for one more minute. Remove the pan from the heat and stir in the parmesan cheese. Allow the stuffing to cool for ten minutes before spreading it on the beef.

With the larger half of steak in the front, overlap the two pieces of steak to create one piece. Cut three pieces of butchers twine long enough to go around the rolled up roast and position them under the steak prior to stuffing.

Spread the stuffing mixture on roast. Leave an inch on each side and a couple of inches at the top of the roast to allow the stuffing to spread evenly without falling out as the roast is rolled. Roll the roast up like a jelly roll and tie with butcher's string to hold. An easy knot is to loop the string three times, pull each end tight, and complete the knot with one loop on top.

Rub the outside of the roast with olive oil and season generously with salt and pepper. Roast uncovered, mushroom-side down in a 350 degree oven for 45 to 50 minutes, 140 degrees is the finished temperature. Pull the roast at 130 degrees. Rest roast for ten minutes before slicing.

step 1
step 2
step 3

POT ROAST (LIKE GRANDMA USED TO MAKE)

Nothing smells better than a nice pot roast cooking on a rainy Sunday afternoon. The aromas remind many of us of family gatherings with good old-fashioned home cooking. Pot roast is one of the easiest and most satisfying dinners to prepare.

3 to 4 pound boneless chuck roast – trimmed of excess fat.

Olive oil– Use to coat all sides of the roast and a couple of tablespoons for searing.

Kosher salt, fresh ground black pepper, granulated garlic

1 tablespoon dried dill

1 tablespoon of fresh thyme – Use ½ tablespoon if using dried thyme**.**

Low salt beef stock – I use the beef stock in a 32 ounce box. For cooking you need enough to braise the beef and the rest is used for gravy.

Rub the roast on all sides with a thin coating of oil. Season the roast well on both sides with the salt, pepper, and garlic.

Using a deep pot, cast iron is best, heat the oil over medium-high heat. Sear the roast for five to seven minutes on both sides.

Add the dill and thyme to the top of the roast. Add enough stock to come up about ½ of the sides of the roast. Bring the roast up to a simmer before putting it in the oven.

Roast covered in a 325 degree oven for three to four hours. Check occasionally, and add stock if needed. The roast will be "fall apart" tender when done.

To make gravy, melt 3 tablespoons of unsalted butter in a sauce pan over medium heat. Stir in 3 tablespoons of flour. Cook for two to three minutes to lose the uncooked flour taste. Stir in the remaining beef stock. Strain the juices from the roast. Remove as much fat as possible with a spoon from the top. Add the sauce to the gravy. Bring the gravy to a boil to thicken. Taste the gravy and adjust the seasoning with salt and pepper before serving.

OVEN ROASTED BABY RED POTATOES AND CARROTS

Oven roasted baby red potatoes and carrots are a natural side dish for pot roast and gravy. Their caramelized sweetness adds a nice complement to the savory saltiness of the pot roast and gravy. The following recipe serves 4 generously.

2 pounds of baby red potatoes

3 medium carrots

1 tablespoon of olive oil

1 tablespoon of kosher salt

½ tablespoon of fresh ground black pepper

1 tablespoon of fresh thyme

Scrub the potatoes and cut them in half.

Peel the carrots and cut them into one inch pieces. Split the thicker large end of the carrots in half again.

In a large bowl, toss the potatoes and the carrots with the olive oil to coat well.

Add the salt, pepper, and thyme and toss to coat evenly.

Line a baking tray with parchment paper to prevent the veggies from sticking.

Spread the potato and carrot mixture over the baking pan.

Bake in the oven for the last hour of the roast cooking time. They will caramelize better if they are on a rack under the roast.

If you are roasting the veggies as a separate side dish, bake in a pre-heated 400 degree oven for 35 minutes.

Using a spatula, flip the veggies over halfway through the cooking process.

Add a splash of olive oil before serving. Garnish with a sprig or two of fresh thyme.

GROUND BEEF

GROUND beef is easily the most popular beef product. When I started in the meat business our most popular ground beef was regular ground beef. It was cheap and contained 30% or more fat. For the more discerning customers we offered a lean ground beef with less than 20% fat. All of our ground beef was produced in-store. Ground beef was made from the trim generated by cutting whole quarters of beef.

As the demand for ground beef grew, we saved the excess fat and used frozen imported lean beef to produce more ground beef. We also ground the leaner frozen beef to produce an extra lean ground beef product.

The commodity name for the frozen beef was boneless bull. It came in sixty pound boxes from Argentina. It was the duty of the meat cutter working the late shift to cut the frozen beef into pieces that would temper overnight to be ground the next morning. It was usually the last thing that was cut on the saw. Yes, the same saw that was used to cut everything from beef, to pork and poultry during the day. I am so happy that food safety is a much bigger issue today. It did not seem wrong at the time because everyone did it. The USDA today requires that retailers state the country of origin on all fresh meat products. If there is imported meat in ground beef it has to say so on the label.

The next evolution of ground beef was the ability to buy pre-fat tested tubes of coarse ground beef. The beef packers boned out leaner ungraded beef and used it to produce tubes of coarse ground beef. The butchers at retail level then ground it through a finer plate to make ground beef. This is still a popular method. Retailers can now buy a variety of fat contents as well as natural and organic choices.

Today's food safety savvy retailer keeps a grind log to track the production of ground beef. Store produced beef trim is ground separate from any tube product. Tracking information from the tube ground beef is recorded as it is ground. It is important for the liability of the product that the retailer know where, when, and how much in the event of a food recall or other food safety issue.

I would not hesitate to ask your local butcher what his program is when producing ground beef. Does he keep a grind log? Does he include unsold cuts from the meat case in his ground beef? Does he regrind unsold ground beef to mix with fresh ground beef? Does he fat test every batch of ground beef?

The hardest question most butchers get about their ground beef is "why is my ground beef dark in the center?" Even the freshest ground beef can be dark in the center and bright red on the outside. I have had customers in the past suggest that we took the time to pack bright red meat around old dark meat just to sell it.

CAUTION: Once you experience your own ground beef, you may never buy store bought ground beef again.

Nothing could be further from the truth. The ground beef on the outside turns red as it reacts to oxygen. The beef in the center is not exposed to oxygen and is naturally darker. The leaner the ground beef the more likely it is to be darker. In some fatter grinds, the fat separates the lean pieces allowing oxygen to react with the meat. The myoglobin in the meat reacts with oxygen to turn the meat red.

The regulation of the sale of ground beef varies from area to area. State and local authorities set the boundaries of product description. The most common way to sell ground beef is by statement of the fat content. The lean percentage is usually stated first, followed by the fat percentage. 80/20 ground beef will contain 80% lean meat with 20% fat. This means that the ground beef can contain no more than 20% fat.

Some areas of the county sell ground beef by description. 80/20 ground beef may be sold as ground chuck. A a leaner 90/10 may be called ground sirloin. In some areas you can only label ground beef, ground chuck if it contains exclusively cuts from the chuck of the beef. The same with ground sirloin, it must be ground only from cuts from the sirloin of the beef.

If all this isn't confusing enough, add in all the other ground beef choices. We can now choose from fat percentages as low as three percent, as well as natural and organic choices. You can also choose cuts such as boneless chuck roast from the meat case and ask your butcher to grind them for you. Keep in mind that if the butcher uses the large grinder that he uses for his everyday production, you may not get the entire roast back. The head of a larger grinder has about a pound of meat that does not get forced out with each grind. Meat is usually run through the grinder twice. A good butcher will catch the first handful of grind to run though at the end so that you get most of what you bought back.

Fat content is the most deciding factor when choosing ground beef for your recipe. 80/20 or ground chuck is a good choice for barbecue patties or meat loaf. The fat will help to hold the patties together when cooking. Choose a leaner ground beef for meat sauces or sautéing. All store-bought ground beef needs to be cooked to an internal temperature of 160 degrees.

Much of the ground beef sold today comes from centralized cutting plants. One of the most common ways to package ground meat products from these plants is to use Modified Atmosphere Packaging or MAP. Some brand-name natural and organic beef producers sell their ground beef in MAP.

Another way stores receive pre-packaged ground meats is conventionally packaged products shipped to the store in a gas flushed pillow-pack. Several packages are placed in a plastic bag or pillow with the oxygen flushed out and replaced by nitrogen. When the bag is opened the product is then scaled with a normal two to three day shelf life.

Most conventionally packaged ground beef has a shelf life of two to three days from the date of process. If you do not plan to use it within that time I would freeze it. MAP is nice because it sometimes has a longer fresh shelf life.

"Pink Slime"

"Pink slime" is a name some in the media have given to what the meat industry calls "lean finely textured beef". It is used as filler in the manufacture of ground beef by packing plants.

The lean finely textured beef is made by heating and spinning beef trimmings to remove the fat. The meat is then treated with ammonia gas and used to make a leaner ground beef.

This manufactured product passes USDA food standards. It is not listed on ground beef as an ingredient because it is just beef. Any ingredients other than ground beef have to be listed as an additive.

There are "pros" and "cons" to this protein extender. The biggest "pro" is use of this product enhances the producer's profit margin and lowers the cost of ground beef. Another benefit is fewer animals are harvested to make ground beef.

The biggest "con" is obviously the unappetizing way the product is made and presented. "Would you like cheese on your pink slime burger?" If you think about it aren't you adding "yellow slime" when you add an egg to your meat loaf mix? One is a natural protein extender and one is a manufactured product.

A Better Idea, Grind Your Own Ground Beef

My personal choice for ground beef is to grind it myself. I started with a small grinder that attaches to my large stand mixer. It worked well for my needs. I recently replaced it with a small freestanding grinder. The freestanding grinder has a larger grinding capacity and also works as a sausage stuffer.

I do not make the choice to grind my own ground beef because of any fear or comfort level with buying store-ground beef. I grind my own just because it gives me exactly the ground beef I want and because it is an economical choice.

Did you ever notice that homemade ice cream always tastes better than even the best store-bought ice cream? The other benefit of grinding my own ground beef is that I feel comfortable cooking patties to a lower internal temperature. I prefer my hamburger patties with just a little pink in the center.

I use a mixture of well marbled tri-tip steak and leaner ball tip steak to grind for my steak burger recipe. Neither cut is a "melt in your mouth tender"; however when ground together they make a succulent and tender burger.

Compare the cost of beef cuts you might choose to grind yourself to the cost of pre-ground beef. Many times you can make better ground beef for less than the cost of store-ground beef. Beef chuck cuts in the summertime are a good example. Their natural fat-to-lean ratio makes them a good choice to grind. Once you grind your own, store-bought grind will never be the same.

Grinding Basics

Trim off any sinew or gristle and leave most of the hard fat on when cutting meat to grind. Butchers do not take the time to remove all of the excess sinew and gristle when generating trim for ground beef. This is another reason that home ground beef is a superior product. You can remove any excess gristle or sinew.

An 80/20 ratio of lean-to-fat is the most flavorful. Leave some fat on the meat you grind. If the ground beef is too lean it will not hold together as a patty or a meatloaf. However, lean cuts from the round work well in dishes such as spaghetti sauce.

Keeping the beef cold is a key to successful grinding at home. Cut the beef in pieces small enough to go through the grinder easily.

Place them on a baking sheet in the freezer for 20 minutes before grinding. The chilled meat will go through the grinder more easily, and will be less affected by the heat of the grinding process.

Grinder heads offer a choice of grinding plates. Most come with two choices, a fine grind and a coarse grind. The fine grind has smaller holes to push the meat through and the coarse grind plate has larger holes.

In meat markets the ground beef is run once through the larger plate. The coarse grind is then ground again through the fine plate. Between grinds, the butcher cleans the sinew and gristle that wraps around the knife blade during the coarse grind process.

Because the holes on the home coarse grinder are much smaller than the holes on the commercial coarse grinding plate, I prefer to grind my home grind twice through the coarse plate. I still remove the blade between grinds and clean the knife blade before regrinding.

A word of caution, once you experience home ground beef, you may never want to buy store-bought ground beef again. A tough chuck steak ad in the hot summer months will have new appeal. "Let's make gourmet burgers!"

FRESH GROUND BEEF STEAK BURGERS

You have never had burgers as good as these. Made from fresh beef steak that you grind yourself, you can grill them to the same medium-rare that you like your steak cooked. The cost of the steak cut used is not much more than any premium store-bought ground beef.

1 ½ pounds of well marbled tri-tip steak

1 pound of ball tip steak (petite sirloin)

2 tablespoons of olive oil

Kosher salt, fresh ground black pepper to taste

Cut the steak into ¾ inch cubes. Leave any hard fat on the steak and remove any gristle. The ball tip steak has a couple of thin gristle strips running through it. The tri-tip is pretty gristle free. The tri-tip may have some soft skin like fat on the outside. Remove it as well.

Spread the cubes on a baking sheet and put the pan in the freezer for 20 minutes. You want the meat to be chilled but not hard frozen.

Mix the two steak cuts together before grinding. After the steak is ground you want to minimize how much you handle it. If using the food processer, divide the steak into three batches before pulsing it into ground beef.

There are a couple of good ways to grind the steak. If you have a grinder attachment for a large stand mixer, grind the meat twice through the larger grinding plate. If not, a gentle touch on a food processer works well. Pulse the meat on short bursts until it is chopped to ground meat consistency.

Divide the ground steak into six equal portions. Shape each portion into a slightly larger patty than the burger bun you are using. This is a premium burger so choose a premium bun.

Line the baking sheet with parchment paper and arrange the patties in a single layer. Press the center couple of inches of each patty so that it is thinner than the outside edge. This will keep the patty a consistent thickness when cooked. Refrigerate the patties for one hour before grilling.

Brush the chilled patties with olive oil on each side and season generously with salt and pepper. Grill on a hot open grill. If you are cooking indoors use a lightly oiled cast iron griddle or pan. Resist the urge to move the patties about when they are cooking. You want to get a nice crust on one side before you turn them over.

Condiments and cheese are of your choosing. For me, the meat is the star of the show. I like it served just burger and bun. Maybe add a little thin sliced red onion and homemade ketchup.

BEEF CUTS FOR SOUPS AND STEWS

THE good news about beef cuts used for soups and stews is that they are all relatively inexpensive. They include most all of the tougher, more flavorful cuts with lots of connective tissue that melts in the cooking process. You can add additional flavor to these cuts by browning these cuts at the start of the cooking process. Roast cuts with a lot of bone such as soup bones or oxtail in the oven to add a nice caramelized flavor to the finished recipe. Brown more meaty cuts such as beef stew meat or short ribs in oil for the same added flavor.

Beef Stew Meat

Beef for stewing is produced from smaller lean pieces of trim that are sometimes included in the ground beef trim. Stew meat usually sells for more per pound than ground beef. A good butcher is proficient at gleaning out the leaner pieces of trim and packaging them for sale as stew meat.

For the most part, stew meat is produced from cuts from the chuck or the round. I prefer stew meat cut from the chuck. It has more internal fat which translates to more flavor.

The most economical way to purchase stew meat is to buy a small roast or steak and cut it into stew meat at home. Boneless chuck steak or roast on sale can be as little as half the cost of packaged stew meat. Choose a lean cut and follow the natural seams to cut your own stew meat. It is a great way to practice your knife skills. You virtually cannot mess up a beef roast cut into smaller stew pieces.

If your budget is unlimited you can choose any cut of meat to make a stew. There are many recipes such as Beef Burgundy or Beef Bourguignon that call for a tender steak cut. The cooking process is much shorter for the meat because it is already tender.

I once had a customer bring me a recipe for a beef stew using beef tenderloin as the meat. To encourage me to try the recipe this special customer surprised me with a purchased package of beef tenderloin from my meat case. How could I not try the recipe with free meat? It basically involved making a nice vegetable stew with lots of fresh herbs. The cubed pieces of steak were then pan seared to a nice medium-rare and folded in just before serving. It was a wonderful take on traditional slow cooked beef stew.

Beef Short Ribs

Beef short ribs are cut from the beef plate. The beef plate is a thick meaty section of ribs from the shoulder of the beef. The butcher slices it into 1 ½ to 2 inch thick slices on the meat saw and then cuts between the bones to produce short rib cuts. One end of the beef plate has more meat on it. The other end has less meat-to-bone. Most butchers package the pieces with a mixture of the meaty and less meat pieces. Look for packages that have more meaty pieces.

Short ribs are a wonderful braised dish. Pan-sear the short ribs then finish cooking in the oven with an inch or so of liquid in the bottom of the pan. Low and slow is the key to a tender dish where the meat just falls off of the bone.

The beef plate is sometimes sliced into ¼ to ½ inch thick slices and sold as kalbi cut short ribs. The most popular way to cook these cuts is to marinate them in a sweet kalbi sauce, then grill them. It is a tasty, economical grilling choice.

Beef Shanks

Beef shanks are the bottom of the front leg of the beef. The butcher slices the shank into one to two inch thick slices. The less meaty slices of the shank are sold as soup bones. The thick connective tissue in the beef shank adds a natural thickener to soups and stews as it breaks down with slow cooking.

The demand for beef soup bones is greater than the demand for shanks. Some butchers trim most of the meat off of the bone to use as ground beef and cut the whole shank into soup bones. Soup bones are popular for making beef stock. Roasting the bones in the oven before making stock gives the beef stock a nice caramel color and flavor.

Some dog owners buy soup bones as treats for their pets. With boxed beef, the good old days when you could ask your butcher for a free dog bone are gone. Butchers buy boxes of frozen femur to sell as pet bones in many stores.

Oxtail

Oxtail is really beef tail. Somehow oxtail sounds more palatable than beef tail. The whole tail is easily cut into smaller pieces with a sharp knife by slicing between the joints. The soft cartilage between the joints and the marrow in the bones add a natural thickener and flavor to soups and stews in much the same way that beef shanks do. Once "poor folks" food, oxtail soup has gourmet status today.

Beef Offal

Beef offal is the generic term given to hearts, liver, tongue, and other assorted byproducts of beef production. Pronounced "awful", it is an appropriate name to me. While I have many friends and family that enjoy these products, I am not so adventurous.

Freshness is the key to buying offal. None of these products benefit from the ageing process. Retailers receive most of these products vacuum-packed in Cryovac®. Some are vacuum-packed in ready to sell packaging.

Some products such as beef tripe come into the store as frozen product. The retailer thaws the tripe and repackages it in smaller packages. Many local health agencies require that previously frozen products be labeled as **previously frozen** when they are displayed for sale.

The best place to shop for offal products is to shop in a retailer that displays and sells a larger quantity of these products. The prices are more apt to be lower and the quality higher in these stores.

BEEF STEW WITH OVEN ROASTED VEGGIES

The veggies for this stew are oven roasted separately and added at the end of the cooking process. The end result is a stew with more depth of flavor. Cutting your own stew meat from chuck is the only way to go. Pre-packaged supermarket stew meat can come from a variety of sources, including the less flavorful round cuts.

2 pound boneless beef chuck roast

Kosher salt and fresh cracked black pepper

Flour (coat beef before browning) – Plus 3 tablespoons for making the thickening rue.

3 tablespoons of olive oil – Plus an additional 2 tablespoons as needed.

Medium onion – diced.

4 garlic cloves – coarsely chopped.

One cup of red wine

24 ounce box of low sodium beef stock

Small bundle of fresh thyme – Tied with a string for easy removal. Plus one tablespoon minced.

2 pounds of Yukon gold potatoes – skin on and cut into bite size pieces.

2 large carrots – peeled and cut into bite size slices.

1 cup of button mushrooms – sliced.

4 tablespoons of unsalted butter

Following the natural seams in the chuck roast, remove excess fat and cut the beef into 1 ½ to 2 inch size pieces. (This is a knife-and-fork stew.) Season the beef generously with salt and pepper. Lightly coat the meat with flour.

Preheat a cast iron Dutch oven pan over medium heat. Add the olive oil and brown the meat in batches. Be careful not to crowd the meat so that it browns evenly. Remove the meat to a small bowl.

Add the onion to the pan and sauté for five to seven minutes, until the onions start to become translucent. Add more oil if necessary. Add the chopped garlic and cook for one more minute. Add the wine, stock, and thyme bundle. Return the beef and any accumulated juices. Stir with a spoon to remove the fond on the bottom of the pan. Bring to a simmer and cover. Cook over low heat for 90 minutes to two hours, (until the beef is tender).

Add the potatoes and carrots to a large bowl. Add enough olive oil to coat the veggies. Season the veggies generously with salt, pepper, and a tablespoon of chopped thyme leaves.

Spread the veggies on a baking sheet lined with parchment paper. Roast for 30 minutes in a preheated 400 degree oven. Pull the pan about halfway through the roasting process and turn the veggies over for more even browning. Wait to start the veggies until the beef has cooked for one hour.

While veggies roast, melt 2 tablespoons of butter and 1 tablespoon of oil in a small saucepan. Sauté the mushrooms for five minutes over medium-high heat.

Remove the mushrooms to a small bowl and melt 2 tablespoons of butter in the saucepan over medium heat. Add 3 tablespoons of flour and stir for a minute or so. Add the rue to the meat to thicken the gravy.

Combine the beef and veggies and serve with crusty bread and a fresh green salad.

Step one: Start with a lean boneless cut of chuck roast

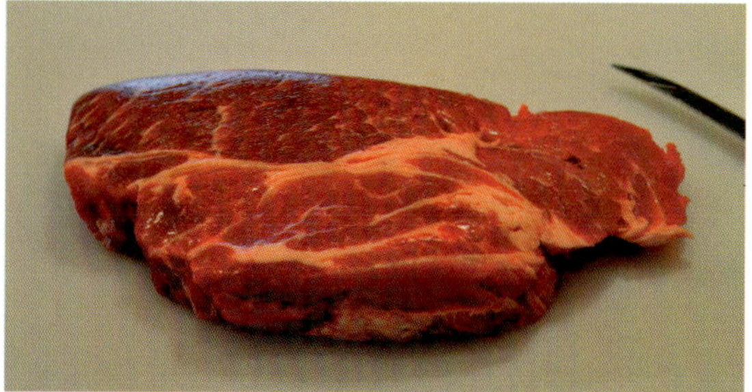

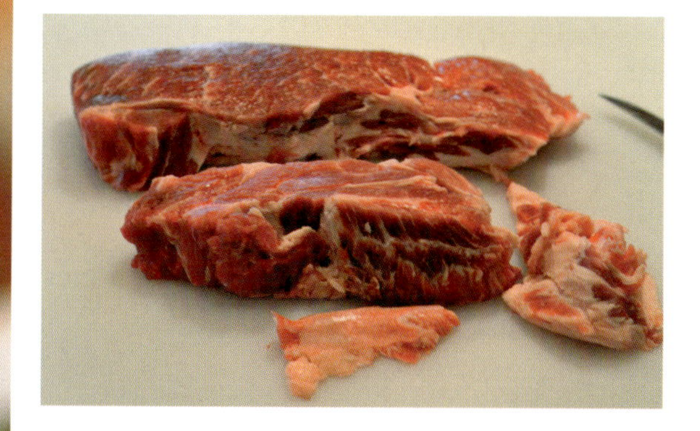

Step two: Follow the natural seam in the roast to remove visible fat or gristle.

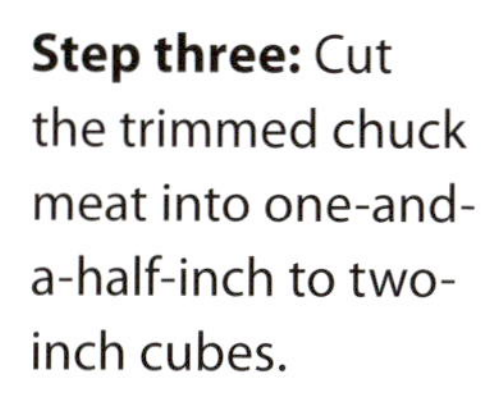

Step three: Cut the trimmed chuck meat into one-and-a-half-inch to two-inch cubes.

Step four: Season the stew meat with salt and pepper. Dust the stew meat with flour. **Helpful hint: A cheese shaker filled with flour works well.**

FRESH PORK

BUYING fresh pork is much easier than choosing the right cut of beef. Through selective breeding and feeding practices the pork sold today is truly the "other white meat". Pork sold in stores today is a consistent lean, more desirable lighter color meat.

All pork produced for retail sale is produced in USDA inspected plants. Unlike beef, the USDA does not have a grading system for fresh pork. Pork is separated by size. The largest demand is for pork that is about 250 pounds and six months old at harvest time.

The biggest choice consumers have in choosing fresh pork products is the choice between pumped pork and non-pumped pork. Some pork producers inject brine into the pork to add flavor and moisture to the cooking process. Other choices consumers have are to choose natural organic pork or to choose heirloom pork such as the Berkshire pork.

Pumped Pork

Pumped pork is sometimes called enhanced pork. The pork producer uses a series of small needles to inject a brine of water, salt, and sodium phosphate. The salt adds flavor and helps to keep the water and moisture in the pork as it cooks. The sodium phosphate helps to keep the moisture in, as well as extending the shelf life of the pork. Some producers also add seasonings to the injected mixture to produce a ready to cook pre-seasoned pork product.

Enhanced pork evolved because of the very lean pork sold in retail stores today. Many consumers judged the pork dry and flavorless as they overcooked it. Enhanced pork makes it easier for home cooks to prepare juicy moist pork. The added moisture makes it more difficult to overcook.

Enhanced pork is often sold as brand name pork. It is against the rules of the USDA to add any products to fresh meat without stating them on the label. Look for ingredients added on the label of brand name pork to see if you are buying enhanced pork. Ingredients on any product are always listed in descending order of how much of each ingredient is in the product.

Natural Fresh Pork

Natural pork should not be confused with organic pork. According to the USDA the term "natural" means that the product has been minimally processed with no added ingredients. The standards for raising and selling organic pork are more involved than producing natural pork.

Natural pork is my personal choice when buying pork. Organic and heirloom pork are both wonderful products, however their cost makes them a special occasion purchase. Natural pork is readily available at competitive prices. Organic and heirloom pork are most often sold at upscale or specialty stores.

Cooking natural pork to a nice juicy doneness requires a little more attention to temperatures and cooking times. Some folks still have a fear of trichinosis and continue to cook pork to a point well past done. With today's feeding practices, trichinosis is not an issue in commercially raised pork. The other point is that trichinosis is killed in the cooking process at 137 degrees.

When I cook lean cuts such as bone-in and boneless chops and roasts, I remove them from the heat when the internal temperature hits 140 degrees. As they rest, the temperature will rise to 145 degrees or more. The end result is a most juicy chop or lean roast. Temperature is usually not an issue when cooking less tender cuts because they have more internal fat and are slow roasted to a juicy doneness.

There are many recipes that involve marinating pork in brine made with water, salt, and sugar. This process adds flavor and moisture in much the same way that pumped pork does. To me, the taste of enhanced pork is not as good as brined natural pork. I prefer to add my own seasonings and flavors to pork when I cook it.

Natural pork is also marketed as brand name pork. The difference is that the label will not list any added ingredients. Some producers market both enhanced pork and natural pork with a similar label. Again, the key is to look for added ingredients.

Another benefit to buying natural pork over enhanced pork is the solution added to the enhanced pork adds weight to the product. When you buy enhanced pork you are also paying for the added weight of the salt and water solution. The percentage of added weight is listed on the label on most enhanced products.

Organic Fresh Pork

Like the rest of the organic meat products, organic pork is raised and produced under strict USDA guidelines. All of the **never ever** standards apply. The pork is never ever given antibiotics or growth hormones. They are fed a diet of 100% organic feed. In addition, the pigs are allowed to roam freely--as opposed to the more confined manner in which commercial pork is raised.

Most organic pork producers start with using heirloom breed pork such as Berkshire. This is not "the other white" meat of commercially raised pork. The growers don't target the size and fat guidelines of everyday supermarket pork. The meat is redder in color, and a more flavorful product.

There are many compelling reasons to eat only organic pork. It is raised more humanely and the meat tastes better. However, it is not readily available and the cost is more than many are willing to pay. Shopping online is a good way to find the organic pork that fits your needs.

Another way to feed your family only organic pork is to buy a whole or half pig from a local farmer. You will have to do your own due diligence to make sure the hog was raised in an organic manner. Some of these operations will also cure and smoke some cuts as part of the process.

Heirloom Fresh Pork

Heirloom fresh pork is produced from hogs with a documented heritage. It is pork from select breeds that have long been recognized as superior pork. The "all white" commodity pork that most of us eat is produced by selective breeding of different breeds of pork. Heirloom pork is produced from a single breed of pork.

Berkshire pork is the one of the most recognized breed of heirloom pork. Berkshire pork is the top of the line in premium pork cuts. If it is not raised organically, it is raised with high standards of feed and the environment in which it is raised. Berkshire is used in the brand name of many premium pork producers.

The meat from heirloom pork has more internal marbling and may be slightly red in color. The flavor of the meat is noticeably better. As a premium cut it commands a premium price. I would encourage you to try a cut of it if your budget allows.

Shop for Berkshire pork in upscale markets or online. Some restaurants also feature Berkshire pork on the menu.

PORK LOIN

The whole pork loin represents the same section of the hog, as the support muscles in the beef. The pork loin includes all of the more naturally tender cuts of pork. Most of the cuts from the loin are quickly cooked with dry heat, much like the steak section of the beef.

Most of the pork loin is very lean. It is very easy to overcook very lean pork. Covering pork chops when pan-frying them can make the chops chewy. The cover keeps the moisture in the pan and the chops get tougher as they steam.

One way to cook moist, tender, lean pork is to start with thicker cut chops. I worked in a meat market where our customers raved about how much better our pork was than the pork sold in the rest of the chain. The reality was that we sold the same pork as everyone else. We just cut it thicker.

When I first started in the meat business we bought whole untrimmed pork loins to process into chops and roasts. All of the pork loin was cut into bone-in cuts. The pork loins varied in size and color of the meat. This is in sharp contrast to the consistent size and color of today's fresh pork.

Pork producers pre-trim the pork loin and break it down into a variety of products. With the increased demand for baby back ribs and boneless loin cuts, much of the pork loin is now sold as boneless cuts.

The sirloin end is cut off and sold a boneless cut. The tenderloin is removed and sold as a boneless cut. The center chop section is sold as boneless center cut loin. The ribs removed to make the boneless loin are sold as baby back ribs.

Retailers still buy whole pork loins. The most common way to sell a whole pork loin is to package the loin end cuts with the rib end cuts and sell them as assorted chops. The middle of the loin is sold as center cut rib and loin chops. The biggest drawback to buying assorted chops is that the package includes the less desirable sirloin chops. Retailers can also buy just the center bone-in section with the loin end and the rib end removed.

Pork Sirloin

The pork sirloin is the least desirable part of the pork loin. The good news is that it is very lean. The bad news is that it has little flavor and is often very chewy. It is rare to find a butcher who buys pork sirloin in any form.

Some stores sell pork sirloin as a sliced bone-in cut. There are about five slices to a bone-in sirloin. The first two cuts are large with small bones. The next three cuts are 30% to 50% bone. Butchers package the sirloin chops with the larger cuts on top, hiding the poor cuts underneath. Again, the other way bone-in sirloin chops are sold is packaged with the blade cuts and sold as assorted pork chops.

If you buy a bone-in pork sirloin roast, you probably will not buy another. It is a hard to slice and serve, and a waste of time. Fortunately, most stores have abandoned trying to sell bone-in pork sirloin roasts.

The best way to enjoy pork sirloin is when it is run through the cuber several times and made into a pork cube steak. You might be attracted to a nice lean package of boneless pork sirloin at a great price, but I would just walk on by.

Bone-in Center Cut Pork Loin Chops

Bone-in cuts are more flavorful than similar boneless cuts. Pork chops are no exception. There are two kinds of bone-in pork chops. They are pork loin chops and pork rib chops. The loin chops look like little T-bone steaks with a loin side and a tenderloin side.

Center cut pork loin chops are usually priced higher than center rib chops. The loin chop is leaner and includes a bite of tenderloin. As always, prices are driven by supply and demand. The demand for lean loin chops is greater than the fatter rib chops. The fact is, some of the more fat rib chops have more flavor and are more tender than lean loin chops.

The best loin chops, of course, are the ones with larger tenderloins. Most butchers package loin chops with these nicer chops on top, hiding the smaller ones underneath. Buying only the nicer chops can be difficult. Hand-picking chops from a full service meat case, is the easiest way to get the ones you want. Another way is to buy thick chops that are usually packaged flat, one or two chops to a package.

One way to cook moist, tender, lean pork is to start with thicker cut chops. I worked in a meat market where our customers raved about how much better our pork was than the pork sold in the rest of the chain. The reality was that we sold the same pork as everyone else. We just cut it thicker.

Center Cut Pork Rib Chops

Center cut pork rib chops have no tenderloin and a part of the rib bone. The first cut rib chops have one solid muscle just like the loin side of the loin chop. As the butcher slices the rib section of the pork loin, the cuts closer to the rib end of the loin have meat that is two different colors. There is a white part and a more red part. *These two red, less attractive chops are the most tender and flavorful chops on the loin.*

When buying natural pork, choose pork chops that are at least one inch thick. They are easier to cook without over cooking them. If you have a friendly customer service butcher, ask to have them cut from the blade end of the rib. Tell the butcher you want the ugly two red colored rib chops at the end of the center loin.

Boneless Center Cut Pork Chops

The boneless center cut pork loin includes both the meat from the loin chops and the ribs chops. With the bone removed, they are all sold as boneless pork chops. You can still choose the better flavored rib chops by looking for the chops with two colors of meat. Sometimes they are easy to find because they are less attractive than the all white ones, and customers pass over them.

Retailers charge more for thin cut chops. Butchers select the more attractive lean chops to slice thin for a higher retail. Save your money. Cut thicker chops in half to make your own thin cuts. Place the chops on edge, fat-side up. Lining up two or more chops in a row makes it easy to keep them upright when you slice through.

Bone-in Pork Loin Roast

The best bone-in pork loin roast is the rib rack. This is the prime rib of the pork. With the rib bones "frenched", the pork rib roast makes a very elegant special occasion dinner. A frenched rib roast is one where the meat is removed from the end of the bones. The other important part of an easy-to-serve rib roast is that the butcher removes the chine bone deep enough so you may slice between the ribs.

Unlike beef, the number of ribs in a whole pork rib roast can vary. One rib per person is a very generous serving. Skilled butchers tie two large pork rib roasts into a circle to make a pork crown roast. They use a large roast needle to shape the two roasts into a perfect circle. I would bet that most of today's generation of less-skilled butchers would not have a clue how to use a roast needle.

Both the bone-in and the boneless pork loin roast are best when roasted to an internal temperature of 145 degrees. Roast them uncovered at 350 degrees and pull them from the oven at 140 degrees. Overcooking either of these lean roasts will result in a dry, tough product.

Boneless Pork Loin Roast

The boneless pork loin roast is easy to cook and serve to a small family or a large gathering. The boneless loin's consistent size and shape make it easy to slice and serve. It is a nice cut to butterfly into a flat shape. The roast can then be stuffed and rolled like a jelly roll and tied. Two whole roasts can be tied together to serve a large crowd.

I prefer the rib end of the boneless pork loin. The rib end is the fatter end of the whole boneless loin. It also identified by the darker portion of meat on the rib end. Six to eight ounces of uncooked meat is a very generous serving to help judge how large of a roast you need.

Save Money by Buying Whole Boneless Pork Loins

One of the best buys in the supermarket meat case today is whole boneless pork loin offered as an ad feature. My local chain store frequently offers whole boneless pork loins for as little as 50% of the cost of precut chops.

They are usually packaged in the same Cryovac® in which the store receives them. A whole boneless pork loin is about two feet long and three to four inches wide. They are easy for anyone with limited knife skills to slice into chops or boneless pork roast.

Cutting a whole boneless pork loin is a simple slice, trim, and wrap for the freezer operation. Tools needed are a sharp knife, a large cutting board, and the recommended safety cutting glove.

Start by identifying the loin end and the rib end of the loin. The loin end is one solid lean piece and the rib end has more fat and two colors of meat. The rib end makes the better roast if you choose to cut part of the loin into chops with a nice roast as well.

A whole boneless center cut pork loin is about two feet long. Some stores also sell a portion of the whole loin. Look for whole loins. Choose the larger of the loins offered for sale. Larger chops are better for single serve chops and the larger roast will be easier not to overcook.

The loin has a leaner loin end and a fatter rib end. Slice chops from the loin end, leaving the rib end as a roast. Count the number of chops needed for your family's dinner. Stop cutting chops to leave a roast at the rib end. During summer grilling season you can slice the roast down the middle, lengthwise to cook it as a wonderful boneless country-style rib.

My local chain store frequently offers in-the-bag savings on boneless pork loins. The local big box discount store displays them every day for an attractive price. Compare the cost of whole boneless pork loin to the retails for boneless chops. It is an easy decision to stock up and fill your freezer.

Pork Tenderloin

The pork tenderloin is the small round muscle on the inside of the pork loin. The whole pork tenderloin weighs about a pound. It can be cooked whole or sliced into smaller cuts and pan-fried or stir-fried.

Pork tenderloin is one of the most versatile cuts in the pork case. It is lean and tender and doesn't require a lot of cooking time. Whole pork tenderloin can be roasted in 25 minutes or less. Thinner cuts of tenderloin cook very quickly.

The key to cooking pork tenderloin is not to overcook it. When I grill or roast whole tenderloins, I pull them from the heat when the internal temperature reaches 135 degrees. After resting, the pork will reach a temperature of 140 degrees or more. The end result is tender, juicy pork with just a tint of pink to the meat.

Fresh pork tenderloin is also a good economical choice. My local retailer frequently offers whole pork tenderloins as an ad feature. I buy them packaged two tenderloins to a Cryovac® bag. They will keep refrigerated in the bag for two weeks or more. I repackage and freeze the extra tenderloin after opening the bag if I am not going to use it for a couple of days.

The Cryovac®-packaged pork tenderloins freeze well in that bag. The unopened thick plastic Cryovac® bag is a nice airtight container for freezing. To use, simply thaw the package in the refrigerator. If you do not need both tenderloins, the extra one can be refrozen for up to three months.

It is important to remove the thin silverskin from the pork tenderloin before you cook it. If you roast the whole tenderloin, the silverskin will shrink and cause the roast to curl. Removing the silverskin before slicing will result in a more palatable cooked product. Folding and tying over the small end helps to make the tenderloin roast more evenly.

PORK RIBS

Baby Back Pork Ribs

Baby backs are the popular choice among many professional and backyard grillers. Because they are cut from the more tender pork loin, they are more naturally tender than regular pork spare ribs. Baby backs are produced when the ribs are removed to make boneless pork loins. Other advantages to baby back ribs are that they are usually more meaty and easier to cut between the ribs to serve.

Baby back ribs are the only product that I consider buying as a pumped pork product. The extra moisture in the enhanced pork makes them almost fool proof to grill. The keys to cooking terrific baby backs are to remove the inner membrane before cooking and to cook them low and slow. Cook them for 1 ½ to 2 hours at 250 degrees. Removing the membrane aids in the penetration of the seasoning and makes the end product easier to eat.

The membrane is the thin, hard covering on the inside of the rib. It is easy to remove. All you need is a sharp knife and a dry paper towel. First determine the loin end and the blade end of the rib rack. The bones at the rib end are shorter and more flat in shape than the bones at the loin end. The bones at the loin end are round in shape.

Slide the point of the knife along the last blade bone to lift up enough of the membrane to grasp it with the paper towel. Using the paper towel, grip the membrane to simply peel it off. Sometimes the membrane will tear when you peel it off. Just repeat the process to remove the remaining membrane.

Pork Spareribs

Regular pork spareribs are the rest of the rib bones under the baby back ribs. The meaty side of the ribs is removed as a slab to be sold as fresh side pork or more often to be processed into bacon. Whole spareribs are sold by size. Three and down (three pounds or less) is the better choice when shopping for ribs. Larger three to five ribs (ribs three to five pounds) come from larger, older hogs and are usually less tender.

The whole sparerib includes the bony brisket end that is difficult to slice. Some butchers cut through this bone several times to make it easier to cut between the ribs. Ask the butcher to crack the bone on the meat saw if you buy whole sides of pork spareribs.

The best choice when buying spareribs is to buy St. Louis style ribs. These are pork spareribs with the bony brisket piece removed. The end result is a smaller sparerib that is easy to slice between the bones for serving. The more wasteful brisket pieces are sold as rib tips. They are an economical choice for a braised rib to cook alone or to add to a variety of dishes.

All pork spareribs benefit from a long and slow cooking process. Spareribs are terrific cooked in a variety of ways from braising to grilling. Personally, I prefer to do the entire cooking process on the grill. I use indirect heat and a low temperature on a covered grill. You can shorten the time on the grill by roasting the ribs in the oven for a couple of hours at 250 degrees. Some people par-boil the ribs in water to shorten the grilling time. I am sure this process makes the ribs tender, but it seems to me that boiling would also take away a lot of the natural pork flavor.

Country-Style Pork Ribs

Country-style ribs used to be cut only from the blade end of the whole pork loin. Today country-style ribs include those cut from the pork shoulder butt as well. There is not a great deal of difference between the country ribs from the pork loin and the pork butt in terms of tenderness and flavor.

As more retailers continue to buy center cut pork loins, more of the country-style ribs are cut from pork butts than pork loins. The blade end is missing from center cut pork loins. Because the pork butt has only one small blade bone, much of the time cuts from the butt are sold as boneless country ribs.

Some retailers also process other cuts of pork into country-style ribs. Lean pork sirloins are sometimes cut into a lean "gourmet" cut of boneless country-style ribs. Ironically, these high-priced ribs are a poor substitute for the country ribs from the pork butt. The pork butt has much more natural flavor than the lean, tasteless, pork sirloin.

HOW TO REMOVE THE SILVERSKIN OF PORK TENDERLOIN FOR ROASTING

Step one: Slide a sharp knife under the silverskin. With the blade tilted up slightly, slide along under the silverskin to remove. Repeat until all of the silverskin is removed.

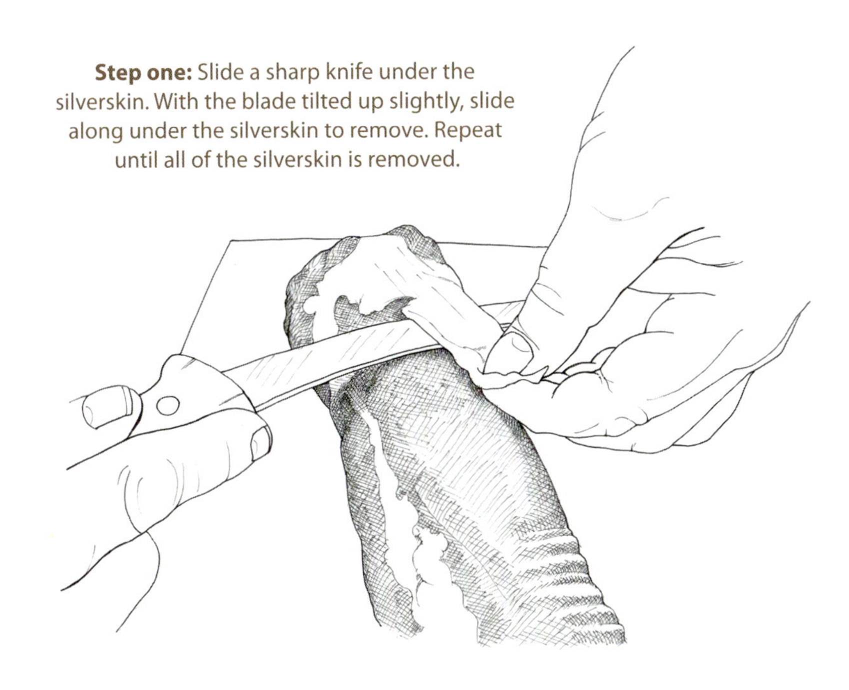

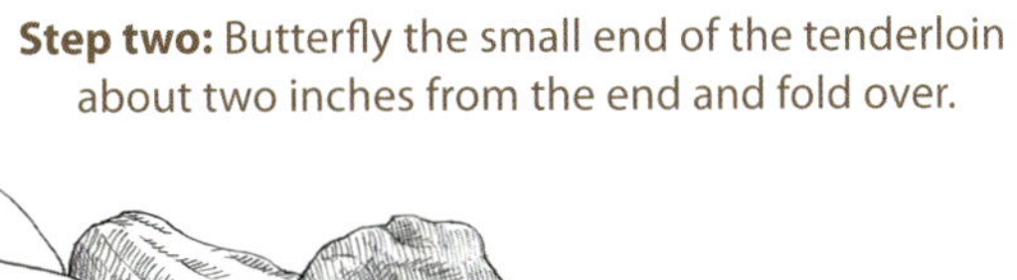

Step two: Butterfly the small end of the tenderloin about two inches from the end and fold over.

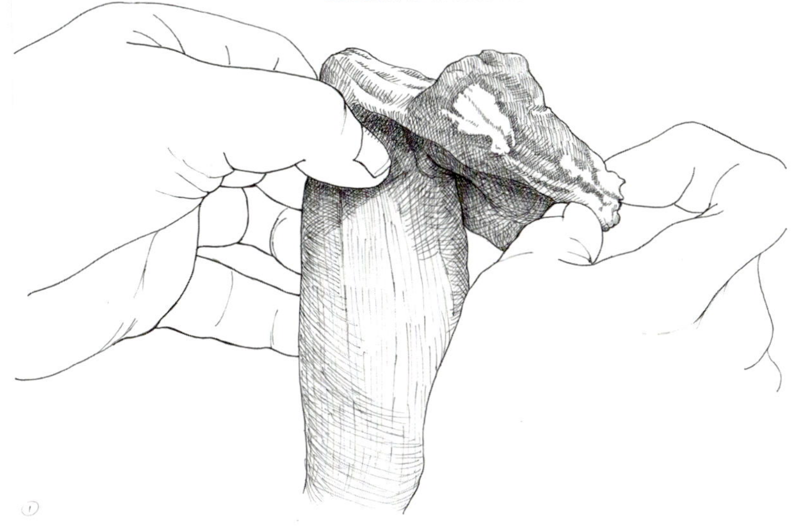

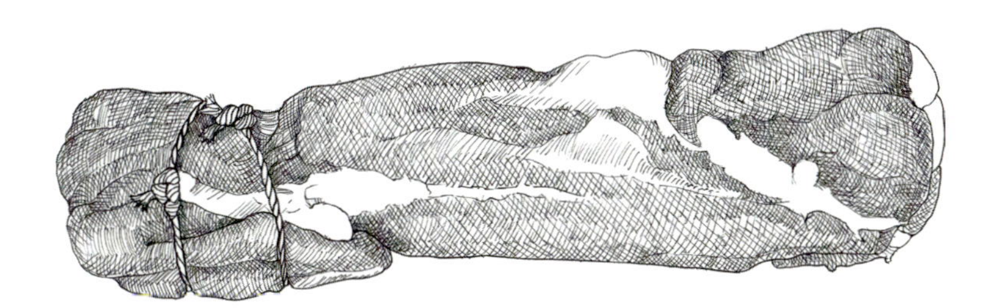

Step three: Tie the folded end to make the tenderloin one consistent shape.

CHIPOTLE RASPBERRY ROASTED PORK TENDERLOIN

This is a quick and easy way to make pork tenderloin into a company dinner. The recipe includes a homemade chipotle raspberry glaze. There are several bottled chipotle raspberry sauces available. Sweet with a little spice, the chipotle glaze sets this dish apart.

2 whole pork tenderloins – Silverskin removed, the thin tail end butterflied and tied back. Two tenderloins will serve six generously.

2 tablespoons of olive oil – Enough to coat the tenderloins

Kosher Salt

Fresh cracked black pepper

Granulated garlic

CHIPOTLE RASPBERRY GLAZE

½ cup of raspberry preserves

¼ cup of water

2 tablespoons of red wine vinegar

1 chipotle pepper – Seed removed and minced, plus one teaspoon of adobo sauce.

Coat the tenderloins with oil. Season the pork generously with salt, pepper, and garlic.

Place the tenderloins in a shallow baking dish and let rest at room temperature for 20 to 30 minutes. Preheat the oven to 375 degrees.

Roast tenderloins in the oven for 20 minutes.

While the tenderloin is roasting, mix the raspberry preserves, water, chipotle, and vinegar in a small saucepan. Stir over medium-low heat until blended.

Remove the tenderloins from the oven. Brush on the glaze.

Return the tenderloins to the oven and roast for five to ten minutes or until the internal temperature reaches 130 to 135 degrees. Pork tenderloin can be very dry if overcooked.

Let the meat rest for ten minutes at room temperature before slicing into ½ inch thick slices. Pour any accumulated juices over the meat and serve.

BUTTERMILK PANKO RIB PORK CHOPS

The spiciness in this dish is softened by the buttermilk and panko coating. It also works well with boneless pork chops or bone-in pork shoulder steaks. The crunchy coating adds a new dimension to fried pork chops.

4 bone-in pork rib chops- Cut one inch thick.

1 cup panko crumbs – You can find them in the Asian section of your store.

½ cup buttermilk

4 tablespoons of olive oil

1 tablespoon Kosher Salt

1 teaspoon of fresh ground pepper

1 tablespoon of smoked paprika – Use regular Hungarian paprika if smoked is unavailable.

½ teaspoon of ground chipotle chili Powder – Substitute cayenne pepper if unavailable.

1 tablespoon of dark brown sugar

Mix salt, pepper, paprika, chili powder, and brown sugar in a small bowl.

Rub both sides with a generous coat of spice rub. Cover the chops and refrigerate for one hour to overnight.

Pour buttermilk into a shallow dish. Spread panko crumbs on a large plate. Dip pork chops into buttermilk and then coat well with panko crumbs.

Heat the olive oil in a large nonstick frying pan over medium heat. Pan-fry the pork chops for six to eight minutes per side until they are done and colored a nice golden brown. Add more oil if necessary when you flip the chops. Remove pork chops from pan and let them rest for five minutes before serving.

Suggested side dishes, oven roasted veggies, rice pilaf.

SPICE RUB BABY BACK RIBS

A generous coating of Jim's Spice Rub and meaty baby back ribs make this dish complete. The sweet honey coating at the end gives the ribs a nice sweet, spicy flavor. Slow cooking on the grill makes them tender and juicy.

LIME AND HONEY GLAZE

2 Tablespoons Lime juice

¼ Cup Honey

Zest of one lime

HONEY AND BALSAMIC VINEGAR GLAZE

2 Tablespoons Balsamic Vinegar

¼ Cup Honey

Peel the membrane from the inside of the rib bones. Use a sharp knife to lift the membrane from the bone on the end. Use a paper towel to grip the peeled up section. It should peel off easily. Removing the membrane improves the rub penetration on the bone side of the ribs.

Rub a thin layer of oil on the ribs. Cover the ribs liberally with rub, cover and refrigerate overnight. Before grilling, let the ribs sit for 30 minutes or so to warm up to room temperature.

If you are using charcoal, arrange the coals for indirect grilling. Place a pan in the center of the grill with the coals around it. For gas grills, turn off the burner under the ribs. Adjust the remaining burner or burners to a lower temperature.

Grill the ribs indirectly on a covered gas or charcoal grill at 225 to 250 degrees for at least one-and-one-half hours. Two hours is better, especially if you are grilling more than one side. Grill the ribs bone side down. Multiple ribs grill well by stacking and then rotating the stacks from top to bottom to keep the heat consistent and the ribs moist. Brown the ribs on both sides before stacking. Rotate them every 20 minutes.

When the ribs are done, lift the lid or remove the cover. Turn up the heat or move the ribs to the hotter section of the grill. Brush the ribs with one or both of the following glazes. Grill for three to five minutes on each side, adding more glaze to get a sticky crust. The glaze recipe is enough for one side of baby backs.

Let ribs rest, covered loosely with aluminum foil for 10 to 15 minutes before serving.

JIM'S SPICE RUB

Try the rub using all of the ingredients. Then feel free to mix and match the optional ingredients to suit your taste. Changing out different chili powders is a fun way to experiment for different flavors.

* ¼ **Cup CALIFORNIA CHILI POWDER.** The first three chili powders are mild chilies. Use one, two, or all three to add layers of chili flavor. I buy them in one ounce packages in the Hispanic section of my store. One ounce is ¼ of a cup. Choose your own combination of mild chili powders.

***¼ cup GROUND PASILLA CHILI POWDER**

***¼ cup NEW MEXICO CHILI POWDER**

¼ cup SMOKY PAPRIKA Substitute regular paprika if you are unable to find smoked.

1/3 cup KOSHER SALT--The flakes of kosher salt melt more quickly.

1/3 cup BROWN SUGAR Run through a strainer if it is lumpy.

2 Tbsp FRESH GROUND BLACK PEPPER – Fresh ground pepper adds more flavor than ground pepper out of a can.

***1Tbsp CAYANNE PEPPER**-- This is the only real heat. Add more or less for your personal taste. Ground chipotle pepper or any other hot chili pepper is also a good choice.

***¼ cup RAW TURBINADO SUGAR** --The natural sugar has a higher burn temperature than white sugar. It doesn't caramelize as quickly.

***¼ cup DRY YELLOW MUSTARD--** Adds a nice tang. Add 2 tablespoons of whole mustard seeds for added pop.

***¼ cup GRANULATED GARLIC**--The granulated garlic mixes better than the powdered garlic.

***¼ cup GRANULATED ONION**--Bigger flakes, like the garlic, mix better than powdered onion.

***1 Tbsp GROUND CUMIN**--Cumin's earthy flavor goes very well with the chili spices.

***2 Tbsp CELERY SEED--** Small whole spice adds an interesting familiar flavor to the rub.

***2 Tbsp DRIED OREGANO--** Rub between your hands as you add it to release the natural oils.

***2 Tbsp DRIED THYME--** Rub between your hands as you add it to release the natural oils.

Mix everything together in a bowl with a large spoon. Pour the rub into a plastic container with a snap on lid. Turn the container upside down and shake well to mix. Fill a shaker and store the rest in the container. The spice rub can be used on any grilled meats or veggies. Use a thin coating of oil to help the rub stick. I buy my spices in the bulk section to save money. I hope you enjoy making your own signature rub, or steal mine and call it your own.

GRILLED CORN SALAD

This is an easy side dish to make for any barbecue. The sweetness of the dish adds a nice contrast to grilled pork.

4 ears of fresh corn

1 medium poblano chili

2 tablespoons of finely diced red onion

2 tablespoons of honey

Olive oil – 2 tablespoons to add to the finished salad and enough to coat the corn before grilling

3 tablespoons of Jim's spice rub or the following rub

1 tablespoon of smoked paprika

2 teaspoons of kosher salt

1 teaspoon of fresh-cracked black pepper

2 teaspoons of dark brown sugar

½ teaspoon of cayenne pepper

Light the coals or preheat your gas grill. Remove the silk and the husk from the corn. Cut both ends of the corn flat when doing so. Place the corn on a baking sheet and coat with the oil. Mix the rub ingredients together in a small bowl. Sprinkle the rub on the oiled corn.

Place the corn and the poblano chili on the grill. Grill the pepper until the skin is charred and blistered on all sides. Remove the pepper to a small bowl and cover with plastic wrap. Grill the corn on all sides for eight to ten minutes until it also has a few charred kernels.

On the same baking sheet that you seasoned the corn, stand the ears on end and remove the kernels using a sharp knife. The left over seasoning on the pan will further season the corn kernels. The skin should come off the pepper easily after it rests in the covered bowl for ten minutes or so. Remove the skin and the seeds and cut it into ¼ inch dice.

Combine the corn, pepper, onion, honey, and two tablespoons of oil in a medium bowl. Serve warm or at room temperature.

Left over corn mixture can be mixed with a beaten egg or two and enough rice flour to make patties to fry for breakfast the following morning.

PORK CARNITAS

While not true to good old-fashioned deep-fried carnitas, this recipe takes less time and is full of flavor. Serve the carnitas with fresh tortillas, fresh guacamole, Pico de Gallo, grated queso fresco (fresh Mexican cow cheese), sour cream, chopped lettuce, and cilantro. One pound of pork will serve three to four people.

1 to 2 pounds, pork shoulder cut – cut into one inch thick chunks two by three inches square. Boneless pork country-style ribs are a good choice.

Kosher salt, fresh ground black pepper, granulated garlic

1 jalapeno pepper – seeded and cut into four strips.

½ to1 cup of orange juice – use the refrigerated low pulp kind.

1 to 2 tablespoons of olive oil

Season the pork generously on both sides with the salt, pepper, and garlic.

Heat the oil in a large bottom pan that had a lid over medium heat. Brown the pork on two sides. Work in batches if you need to. Meat does not brown well if the pan is crowded.

Return all of the meat to the pan. Add orange juice until the top pieces are half-covered. Ideally the meat will be in one layer. Add the jalapeno pepper. Bring to a simmer, cover, and reduce the heat to low. Cover and simmer for 45 minutes. Turn the meat over two to three times during the total cooking process.

After 45 minutes, remove the cover and continue cooking for another 45 minutes. Remove the meat to a cutting board. Let it rest.

Reduce the remaining sauce to ¼ to ½ inch. Cut the pork into half inch slices. Toss the pork with the sauce and reheat to serve. Serve the carnitas with fresh tortillas, fresh guacamole, Pico de Gallo, grated queso fresco (fresh Mexican cow cheese), sour cream, chopped lettuce, and cilantro.

PICO DE GALLO

4 Roma tomatoes

1 jalapeno pepper

1 bunch of green onions – sliced thin.

1 tablespoon of chopped cilantro

1 fresh lime – juiced

Combine the ingredients, add salt and pepper to taste.

PORK SHOULDER

The pork shoulder includes the pork picnic cut and the shoulder butt. Sometimes they are sold as one piece, but more often they are sold as separate cuts. The one piece cut is popular as a barbecue restaurant cut, and is often used in professional barbecue competitions.

It is interesting that the pork butt is nowhere near the back of the hog. The name comes from the names used for separating the whole back leg, the butt half and the shank half.

Fresh Pork Picnic

The pork picnic is the front leg of the hog. Most often it is sold as a smoked pork picnic. Although it looks and tastes like ham, the smoked pork picnic cannot be labeled as ham. The USDA designation for ham includes only smoked cuts from the back leg.

As a fresh cut the pork picnic is not a regular item in most pork sets. There is a lot of waste in a fresh picnic, making it a hard sell as an everyday product. It has a large knuckle bone at one end, a thick shank bone at the other end, is covered by a thick layer of fat, and has a thick skin covering.

There is a thick meaty section in the middle that makes all of the loss worthwhile if you enjoy moist, tender pork. A covering of fat and thick skin keep the meat very moist during the longest slow cooking processes. It is easy to see why it is a favorite of barbecue specialists.

Pork Shoulder Butt

The whole pork but is a square shaped piece of meat about four to five inches thick and six to eight pounds in weight. There is only one small bone in the pork butts marketed today. The pork butt has more internal fat than any other cut of pork with the exception of bacon. It is this fat that keeps the meat so juicy and tender when cooked.

There are several ways the butcher can cut the pork butt. Pork steaks are cut from the bone-in end of the butt, with the remaining boneless piece cut into country ribs or tied as a roast. The whole butt can be sliced into 1 ½ inch thick slices that are cut into country ribs. The butt is also sold as a bone-in roast and tied as a boneless roast.

Pork butt cuts benefit from a low and slow cooking process. The natural fat in the meat keeps it moist during cooking. Slow roasted or braised, the pork butt roast is easy to cook to "fall apart tender", (think barbequed pulled pork).

Pork steaks and country-style ribs are tender enough to pan-fry or to cook directly on the grill. Braising or slow roasting these cuts will result in tenderness similar to the roasted pork butt.

My favorite way to cook a nice whole pork butt is to slow smoke it for at least ten hours at 225 degrees. It is important to smoke the roast ten hours or more to make it "fall apart" tender. The meat will be cooked well before, but the additional cooking time finishes the cooking process. I make a paste of my spice rub with oil, and coat the entire roast before placing it in my small outdoor smoker. Resist the urge to trim any fat from the roast. Most of the fat will melt away during cooking and will help to keep the roast moist.

I use an inexpensive Brinkman® charcoal smoker. The smoker is a small cylinder about three feet tall and sixteen inches wide. The meat is placed fat-side up on the top grill. Under the meat is another grill, with a large metal bowl for wood chips and water. The charcoal is burned at the bottom of the grill. There is a small door to feed the fire at the bottom. Using natural hard wood charcoal as opposed to charcoal briquettes seems to work best. It is a great way to spend a summertime family gathering day. Just get up early and feed the fire every hour or so.

The end result is a beautiful dark-crusted pork roast that just falls apart when you cut into it. Serve it shredded, with or without your favorite barbecue sauce. The best part is that you can serve a small crowd a memorable dinner for less than one dollar a serving. If you compare it to steak you may be able to buy the grill for the difference in cost!

Another popular use for pork butts is in the making of fresh and smoked sausage. Many local specialty sausage makers use boneless pork butt as their meat of choice. It has the right amount of fat-to-meat ratio to make wonderful "gourmet" sausages.

Making your own specialty sausage at home is surprisingly easy. The variety of seasonings and secret recipes lend an air of mystery to making sausage. All you need is fresh ground meat and a few simple starter recipes to create your own signature sausage. Pork is the most popular meat for sausage, however you can use any ground meat.

I use the small grinder attachment on my large stand mixer to grind pork, lamb, turkey and chicken to make sausage. If you do not have a grinder, many larger stores sell pre-ground unseasoned meats.

Following are a few basic sausage recipes. Feel free to use different ground meats or to add other fresh ingredients to your own signature sausage. Examples; Add diced apple chunks or other fruits to the breakfast sausage. Add roasted jalapenos or caramelized onions for a more savory flavor. Substitute fresh herbs for dried herbs.

One of the biggest criticisms of fresh sausage is the salt level. The best way to check your seasoning and salt level is to cook a small sample after you mix the ingredients. Fresh sausage benefits from resting overnight in the refrigerator. The time gives the ingredients a chance to blend with the meat.

If you grind your own meat at home, place the meat in the freezer for thirty minutes prior to grinding. Choose fatter cuts such as pork butt, skinless poultry thighs, and lamb shoulder for making sausage.

Saving Money Buying Pork Butts

The pork butt is one of my favorite cuts of pork. It is inexpensive, it is flavorful, and it is versatile. The least expensive way to buy pork butts is to buy them packaged two whole butts to a Cryovac® bag. This is the same way that the retailer receives them to process into steaks, roasts, and country ribs. The cost is sometimes as much as 50% less than the weekly ad feature. I also find them at my large warehouse store as a bulk purchase.

Sometimes the pork butts are sold Cryovaced® with the bone removed. If not, it is a fairly easy process to remove the bone. With the bone removed it is it is one of the most rewarding home butchering projects you can do. One bag of two roasts can yield a variety of roasts, steaks, country ribs, pork stew, and sausage for your freezer. All of this comes at a cost significantly lower than the same cuts purchased in retail packaging.

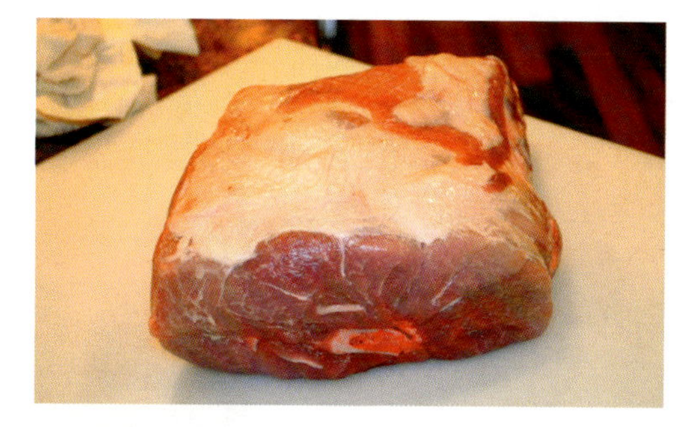

Step one: The whole pork butt has a fat covering on one side; the other side is lean meat.

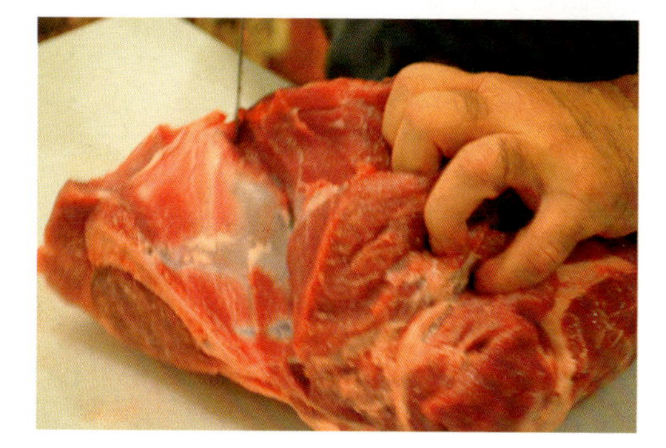

Step two: Turn the butt fat side down to remove the blade bone. The blade bone is flat on the meat side of the pork butt. Using a sharp knife cut along the bone to expose the entire flat side. Using the point of the knife cut along the shape of the bone. *It will help to remove the whole bone more easily when you turn it over and trim along the top part of the blade bone.*

Step three: Note the shape of the top of the blade bone. Turn the pork butt over and following the shape of the bone, remove it.

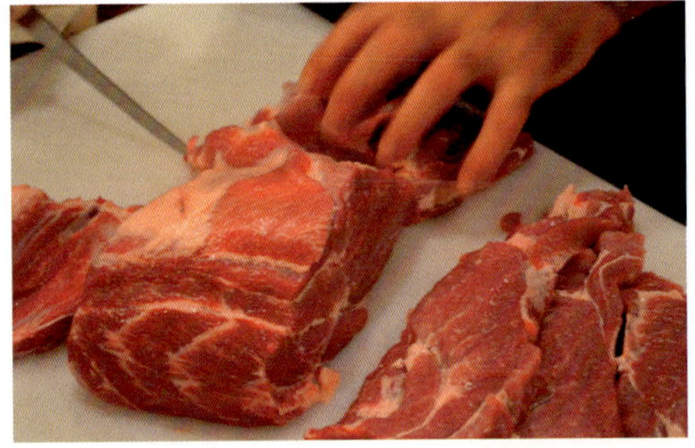

Step four: Remove the loose part of the pork butt. *The loose part is the part that you just removed from the bone.*

Step five: Slice the remaining piece into pork steaks. Leave this piece whole if you would rather have a roast. Cut the loose piece and any larger pieces into one-and-one-half-inch boneless country-style ribs. *Cut smaller pieces into cubes for pork stew or grind for ground pork or sausage.*

Note: The whole pork butt can be tied as a boneless roast if you want a large roast. The whole boneless butt can also be cut into boneless country-style ribs or ground for sausage.

FRESH BREAKFAST SAUSAGE

1 pound of ground pork – Feel free to substitute ground chicken or ground turkey.

1 teaspoon of Kosher salt – start with 1 teaspoon and add more to taste after cooking a sample.

½ teaspoon of fresh cracked black pepper

1 tablespoon of fresh sage leaves – minced fine – substitute 1 teaspoon of dried sage.

¼ teaspoon of cayenne pepper – add just a pinch or none if this sounds like too much to you.

1 teaspoon of fresh grated ginger – add just a pinch if using dried ginger.

Keep the meat chilled and blend in the seasonings. It is important to not over work the meat during mixing. Pan-fry a small sample. Adjust the salt level and refrigerate overnight. Enjoy!!

Suggested additives; 2 tbsp diced apples, pinch of nutmeg, 2 tbsp diced mango, a dash of cold white wine, tbsp of maple syrup (not all at once of course)

FRESH ITALIAN SAUSAGE

One pound of ground pork – substitute ground chicken or ground turkey.

1 teaspoon of Kosher salt – start with 1 teaspoon and add more to taste after cooking a sample.

½ teaspoon of fresh cracked black pepper

½ teaspoon of sugar

1 teaspoon of fennel seeds

¼ teaspoon of caraway seeds

Keep the meat chilled and blend in the seasonings. It is important to not over work the meat during mixing. Pan-fry a small sample. Adjust the salt level and refrigerate overnight. Enjoy!!

To make a hot Italian sausage, add ½ teaspoon or more (to taste) of crushed red pepper flakes. Suggested additives; a tablespoon of roasted red peppers, minced roasted garlic, diced caramelized onions, diced smoked gouda cheese.

FRESH PORK CHORIZO

One pound of ground pork

1 teaspoon of kosher salt

½ teaspoon of fresh cracked black pepper

1 tablespoon of red wine vinegar

1 tablespoon of smoked paprika

1 teaspoon of granulated garlic

1 teaspoon of fresh oregano – minced fine – substitute ½ dried oregano.

½ teaspoon of dried chipotle pepper

Keep the meat chilled and blend in the seasonings. It is important to not over work the meat during mixing. Pan-fry a small sample. Adjust the salt level and refrigerate overnight. Enjoy!! Chorizo is wonderful scrambled with eggs or used as a taco filling.

FRESH PORK BRATWURST

One pound of ground pork

1 teaspoon of kosher salt

½ teaspoon of fresh cracked black pepper

1 teaspoon of fresh grated ginger

½ teaspoon of ground nutmeg

Keep the meat chilled and blend in the seasonings. It is important to not over work the meat during mixing. Pan-fry a small sample. Adjust the salt level and refrigerate overnight.

Shape the sausage into patties grill them as an alternative to ground beef patties.

FRESH PORK

The following pork cuts are most often sold as seasoned and or smoked products. Cooked in their fresh form they are also delicious.

Fresh Leg of Pork

The whole pork leg is the largest of the bone-in fresh pork roasts. A whole pork leg usually weighs between sixteen and twenty pounds. Fresh leg of pork is usually only on display in most retail stores during the Christmas holiday season. Expect to ask for a special order during the rest of the year. Most of the pork legs are cured and smoked to meet the demand for ham and ham products.

If you purchase a fresh leg of pork, ask the butcher to remove the aitch bone (pronounced H-bone). The aitch bone is the hip bone on the butt end of the leg. Removing the aitch bone leaves you with just the long round femur bone to navigate when slicing.

A whole leg of pork is a good choice for feeding a large group for a special occasion dinner. The whole leg works well with a variety of seasonings and coatings. Look for recipes on line to find one that fits your needs.

Slow roast the fresh leg of pork, fat-side up, at 350 degrees until the internal temperature reaches 150 degrees. Depending on size, allow four to five hours for this large roast to bake.

Fresh Pork Belly

Pork belly is the thick, fat slab of meat removed when the producer trims off the sparerib. In raw form it has a covering of skin on one side. Some bacon is produced with the skin left on, while most remove the skin before curing.

The most common use for pork belly is to be cured and smoked into bacon. Fresh pork belly has found recent favor among some upscale restaurants. Chefs fry and braise it to make a variety of fancy dishes.

Uncured fresh pork belly is sometimes sold as sliced fresh side pork. My mom used to season it with salt and pepper, dredge it lightly in flour, and fry it up crispy. I honestly have never cooked side pork, but the memory of my mom's fried side pork makes my mouth water.

Fresh Ground Pork

Fresh unseasoned ground pork is readily available in many supermarket butcher shops. Most local regulators do not require a fat content to be stated on the label. The fat content will vary from source to source.

There are a number of recipes that call for fresh ground pork. It is used in everything from spaghetti sauces to stir-fried Asian dishes. Fresh ground pork is added to other ground meats to make a better meatloaf or meatballs. Any recipe using fresh ground pork should be cooked to an internal temperature of 160 degrees or more.

SMOKED AND CURED PORK

EARLY man quickly discovered that adding salt and smoking meat was a terrific way to extend the shelf life of meats. Some of the best recipes for curing and smoking meat have been around for centuries. Pork has long been the most popular meat ingredient to cure and smoke.

Smoked and cured pork accounts for a large percentage of the pork sold. Virtually every cut of pork is offered in a cured and smoked version. If you don't think the demand for cured and smoked meats is high, just compare the size of the fresh pork display to the ham, bacon and sausage set in your local supermarket.

One interesting note about cured and smoked meats; unlike fresh meats that "bloom" after the newly cut surface is introduced to oxygen; smoked meats turn gray with exposure to light and oxygen.

Vacuum-packaged smoked products retain their bright color until they are opened or lose their seal. There is no immediate loss of quality but the product does not look as appetizing.

You can sometimes save money by buying marked down vacuum-packed smoked products that have lost their seal. The industry appropriately calls these products "leakers". Check the "sell-by" date before buying leakers. Treat the leaker as opened product. Most smoked products are good for a week or so after they are opened.

Smoked Ham

Choosing a smoked ham is almost as difficult as choosing what kind of beef is the best for your needs. There is a huge variety of smoked ham from which to choose. The choices are many, from bone-in and boneless, to dry-cured and wet-cured.

The USDA requires that all smoked meats that are labeled as ham must come from the leg of the pork. Smoked pork shoulders are labeled as smoked picnics. Turkey ham, made from dark turkey meat is always followed by the statement "cured turkey thigh meat".

The other question when buying smoked ham is whether it is ready to eat or if it needs to be cooked. If a ham is ready to eat, it will include the statement "fully cooked" on the label. Many fully cooked hams have directions on the label for reheating. If not, reheat the ham at 325 degrees to an internal temperature of 140 degrees.

Dry-cured Ham

Dry-cured hams are sometimes called country hams or prosciutto. The end result is a salty ham that is either sliced very thin for serving (as in prosciutto) or soaked in water before cooking (as in an old fashioned country ham).

To make dry-cured hams, the processer rubs a salt and spice mixture to the outside of a fresh ham. The curing process can take from six months to a year or more. During the curing process, the salt penetrates the meat and removes much of the moisture. The end product will weigh as much as 25% less than the starting weight. Hams such as prosciutto are just cured, and not smoked. Country hams are usually smoked as part of the process. Follow the directions on the label of country hams for safe preparation and storage.

Wet-cured Ham

Wet-cured hams are the most popular process and are most of the hams sold today. The fresh meat is injected with a curing and flavoring solution before cooking or smoking. In some cases, smoke flavoring is added to shorten the smoking process. Not all smoked meat is smoked using smoldering wood chips. One popular process today is to add atomized smoke flavor during the cooking process.

Hams labeled as hickory smoked, honey cured, sugar cured, or extra lean must meet a minimum USDA standard to do so.

Wet-cured hams are popular because the process shortens the curing and smoking process to produce a consistent, affordable product. In spite of all of the marketing names that allude to an old-fashioned dry-cured ham, virtually all of the hams sold in most supermarkets are wet-cured.

Bone-in Ham

A whole untrimmed pork leg has quite a lot of waste. The leg starts with a thick fat covering, a long shank bone, and the hip bone (called the aitch bone).

Removing the aitch bone makes the butt end of the ham almost boneless. All that is left is the easy-to-navigate femur bone. The leg can be cured and smoked as is, or with all or part of the waste removed.

Most regions of the country have a more expensive premium bone-in ham. Most of these premium hams are made using the wet-cured method. The producers differentiate themselves from the competition with their seasonings and higher trim standards.

They remove the aitch bone, shorten the shank bone, and remove much of the excess fat from the outside of the ham. Buying one of these local favorites would be my first choice if money does not matter.

Many national ham producers also market a similar premium bone-in ham. This would be my next choice if not on a budget. A very popular way to sell bone-in hams is to sell them as a spiral-sliced ham. The ham is placed on a special slicing machine that spins the ham around while slicing it to make an easy-to-serve bone- in ham.

Bone-in ham is sold as whole and half-hams. The more round end of the ham is the butt half. The lower leg part of the ham is the shank half. The butt half usually has more meat and is priced higher than the shank half.

Some supermarkets sell a less expensive bone-in ham that can be priced as a low price "loss leader". These hams are usually poorly trimmed with the excess fat and bone left on. They still have the long shank, and the aitch bone will still be on the butt end.

Ham Portions

Another way producers get the cost down is to sell ham portions. The ham is still cut into butt and shank halves. However, they remove several meaty center slices before packaging.

The center slices are packaged separately and sold at a higher retail. The remaining ham is sold as ham shank portions and butt ham portions. All bone-in hams that have center slices removed must be labeled as ham portions.

Boneless Ham

Boneless ham is a great way to enjoy smoked ham. Like bone-in hams, they are available in a variety of qualities and values. The better boneless hams are made from solid pieces of the pork leg. For the most part, it is a case of you get what you pay for. Better quality boneless hams usually cost two to three times as much as the cheaper hams.

The less expensive versions are sometimes labeled as "ham and water product." These cheaper versions are made from the trimmings and shaped into a loaf. It is easy to tell at first slice whether you have a solid muscle ham or one that has been chopped and formed.

Save Money When Buying Ham

The best time to buy hams is during Easter and Christmas holiday sales. Many supermarkets sell hams at cost or below as a holiday loss leader. The "sell-by" date on most smoked products is sometimes two to three months past the holiday. Buy an extra ham and leave it in the refrigerator to use later. Don't freeze it if you will use it within the "sell-by" date.

Freezing water cured hams is not recommended. The moisture between the meat fibers will make the ham grainy when thawed. Roasting the ham before serving will help to renew the ham.

Some stores have an overstock of hams after a holiday sale. The sale of large hams slows dramatically after the sale is over. Check the "sell-by" date on the sale hams. Go back to the store a week or so before they go out of code. You can sometimes find them on sale at less than the holiday price.

Bacon

I belong to that group of people who think bacon should be one of the major food groups. What can you not do with bacon? It is good for breakfast, it is nice for lunch, and it is wonderful for dinner. My favorite local donut shop even sells a maple bar with a slice of bacon on top. Yummm!

The most readily available bacon is made from cured and smoked pork bellies. The pork belly is the meat on the outside of the pork sparerib. Cured and smoked, it turns into bacon. Old-fashioned cured bacon is cured with a dry rub of salt and seasonings. Most of the bacon sold today is wet-cured in much the same way as ham. The cure is injected into the pork belly for a faster curing process.

Most bacon is sold pre-packaged with a little window to view the how lean or fat the bacon is. Choosing this bacon is usually a combination of lean and price for most shoppers. Most supermarkets feature bacon as part of their weekly ad.

Pre-packaged, sliced bacon freezes well for three months or more. Save money by stocking up if the bacon price is hot. Thaw the bacon in the refrigerator and use it within a couple of weeks.

Gourmet bacon

Gourmet or premium quality bacon is made from hand-selected pork bellies that are usually leaner. They also may be bellies from higher quality heirloom pork. The bacon is sometimes (but not always) dry-cured. Most often it is smoked with real wood smoke.

Like hams, most regions have locally produced premium bacon. If the producer is large enough to have a distribution system, you may be able to find it in your local supermarket.

Other choices to find good bacon are to shop the internet or to source a small local smokehouse. A weekend trip to the country to bring home the bacon is not a bad activity.

My favorite bacon is thick cut pepper-coated bacon sold by the slice in my local supermarket service case. The brand name of the bacon is Hempler; a Seattle based family smoked meat business. It is more expensive than packaged bacon but I rationalize the cost by thinking that eating leaner bacon is a healthy lifestyle choice.

I find that the easiest way to cook thick-cut bacon is to start with a cold frying pan over medium-low heat. Cooking this way seems to minimize the grease splatter and gives you good control over the crispiness of the cooked bacon.

Smoked Pork Jowl

Smoked pork jowl is exactly what it sounds like. It is the jowl of the hog, cured and smoked. Smoked jowl is a smaller fatter version of smoked slab bacon. The jowl is a small fat square of pork about a pound or so.

Smoked jowl is an economical choice to slab bacon. It is usually sold with the rind, or skin left on. If you have a customer service butcher, ask him to remove the rind before slicing it for you. At one end of the sliced jowl is a small dark fat gland. The fat gland can have a bitter taste, so cut it out before cooking.

Pancetta

Pancetta is made using the same fresh pork belly from which smoked bacon is made. Pancetta is cured with salt and spices and does not go through the smoking process like smoked bacon. The most common way to prepare pancetta is to roll the cured pork belly like a jelly roll. The pancetta is then tied into this round shape and hung in a cool dark place to finish the curing process.

The cured pancetta is then sliced and fried for use as an ingredient in cooked dishes. Pancetta, sliced and cubed, can be substituted for fried bacon. Pre-packaged pancetta is available in most large supermarkets. You can find slice-to- order pancetta in many Italian delis and full service meat markets.

Canadian bacon

Canadian bacon is made from smoked boneless pork loins. If you like very lean bacon, Canadian bacon is a good choice. A nice slice of Canadian bacon is the meat of choice in most Egg Benedict dishes.

Canadian bacon is made in much the same way as bacon and ham. The meat is injected with wet cure and smoked. It is most often sold sliced and pre-packaged. Sliced thin it is Canadian bacon. Sliced thick it is sold as smoked boneless pork loin. Smoked boneless ham is a reasonable substitute for Canadian bacon in most recipes.

Smoked Bone-in Pork Loin

Smoked bone-in pork loin is much more flavorful than its boneless pork loin cousin. It is sliced into ¼ to ½ inch thick slices and sold as smoked pork chop. Grilled or pan-fried, it only needs a side dish or a salad to complete a dinner.

Smoked bone-in pork chops used to be more popular than they are today. Slicing smoked bone-in pork loins into chops used to be a regular task in most butcher shops.

Today most smoked bone-in chops are sold pre-packaged in the smoked meat section. I would encourage you to try a smoked bone-in chop if you have never tried one.

Smoked Hocks and Shanks

Smoked hocks and shanks are the smoked lower leg of the hog. Smoked hocks are smaller and have a covering of skin. Shanks are further up on the leg and have a better meat-to-bone ratio than hocks. Smoked shanks are a meaty version of smoked hocks. Smoked hocks and shanks are similar in salty, smoky flavor.

Smoked hocks and shanks are used to add flavor to slow-cooked dishes like beans or soups. Both are usually removed at the end of the cooking process. Choose the smoked shank if you want to add more meat back to the dish.

One of my favorite dishes using smoked hocks or shanks is split-pea soup. It is almost magic how an ugly pot of dried split-peas, chopped onion, hock, covered with water turn into smooth, creamy split-pea soup. Just put all of the ingredients in a large pot cover and let it simmer for an hour or so.

Smoked and Cured Sausage

The variety of smoked and cured sausage available would require another book to cover and there are already several books out there. The time-honored recipes and methods of curing and smoking meats and sausages is called charcuterie. For a long time most of the better cured meats were imported. You had to seek out specialty delis to buy the best in cured meats.

Fortunately, there seems to be a renewed interest in the making of these high quality products. Restaurant chefs are doing their own in-house curing and smoking. People are traveling to Europe to learn old-world charcuterie ways and opening specialty shops back home.

Today there are more home-grown suppliers than ever. You can buy an increasing amount of high quality sliced and cured products pre-packaged in your local supermarket.

Most of the lunchmeats, hot dogs, and smoked sausages sold in today's supermarket are produced in much the same manner as the mass produced hams and bacon. Efficient manufacturing process is used to make a consistent quality product that can be priced competitively.

The day when lunchmeats and hot dogs were made of mystery meats is long gone. I am sure there was a time when you would not want to know the ingredients in your favorite hot dog.

The USDA guidelines today are very clear. Any added ingredient to any meat product must be stated on the label. Ingredients are listed in descending order from most to least. If the hot dog ingredient list states turkey, chicken, pork, the hot dogs contains more turkey than chicken.

Scarier to me than nitrates, sodium erythorbate, sodium phosphate, and potassium chloride, are terms like poultry skin. Poultry skin is left on the poultry to add fat to the finished product. Ground up and emulsified poultry skin, Yuck!! The nitrates and other products are just added to help in the curing and shelf life of the finished hot dog.

If you are concerned about the nitrates used in cured and smoked products you have a couple of choices. There are smoked bacons and hot dogs available from several natural producers. Because these products don't contain nitrates they are most often sold frozen. Keep them frozen until you cook them. The other choice obviously is to not eat cured and smoked meats.

To me, the eating of nitrates and salt used to cure meats is almost as large of a political issue as it is a health issue. Every day the media presents us with a new food we should eat, or not eat, to be healthy. I am not discounting that eating excess salt and nitrates is a very serious health issue. I think the key is excess.

Have you ever noticed that the meat in cabbage rolls is pink? The meat is turned pink by the natural nitrates found in the cabbage. Nitrates are found naturally in the cabbage.

Starting bacon in a cold frying pan over medium-low heat helps to minimize splatter and gives you good control of the final crispiness.

OVEN ROASTED BACON JALAPENO PEPPERS

This is always the first appetizer to go at any summertime barbecue. Who doesn't like bacon? With a limited ingredient list, they are surprisingly easy to make.

12 medium sized jalapeno peppers – seeded and cut in half lengthwise.

12 ounce brick of cream cheese – softened to room temperature.

12 ounce package of thin sliced smoked bacon- warmed to room temperature.

¼ cup of chopped chives for garnish

Preheat the oven to 350 degrees. Line a baking sheet with parchment paper.

Wash the jalapenos. Cut them in half and remove the seeds and inside ribs with a spoon.

Using a butter knife, fill each pepper with cream cheese. Flatten the top to the height of the pepper.

Cut each slice of bacon in half. Tuck one end under the pepper and wrap the stuffed pepper with bacon. The warm bacon will stretch and make this process easier.

Place the peppers on the baking sheet and bake for 30 minutes, until the bacon starts to crisp.

Garnish with the chives and serve.

FRESH LAMB

THE amount of display awarded for fresh lamb in most supermarkets is very small. The reasons are twofold. Fresh lamb is not a widely popular meat to cook at home and the total cuts from a whole lamb don't take up much space.

All lamb sold at retail is inspected by the USDA. Just like beef, the grading of lamb is voluntary. Lamb sold at retail level is graded Prime, Choice, and Good. The grading system uses fat marbling as its major criteria. Most of the lamb sold in retail stores is USDA Choice or Good.

The USDA states that lamb is meat from a sheep that is less than one year old. Most lambs are harvested between 6 and 8 months old. Sheep older than one year are sold as mutton. Mutton has a stronger flavor and is less tender than lamb.

Some lamb producers market their lamb as a brand name. Much of this lamb is delivered to the stores precut and packaged. The cuts are either vacuum-packed or MAP (modified atmosphere packed). The advantage to both the retailer and the consumer is the longer shelf life. Some vacuum-packed product is delivered frozen. The retailer then thaws the product in the cooler before displaying it for sale. It is OK to refreeze this product, but I would ask the butcher for still frozen product if you intend to keep it frozen.

The lamb marketplace includes a large percentage of imported lamb. Australia is the largest exporter of lamb to the USA. There was a time when the imported lamb did not compare well to domestic lamb. It had a stronger flavor that most people did not care for. Today's imported lamb compares favorably to domestic lamb in flavor. As a rule, most cuts of imported lamb are less expensive than domestic lamb. Restaurants are largely responsible for creating a demand for quality imported lamb.

There are only five basic primal cuts of lamb. They are shoulder, shank/breast, rack, loin, and leg. The USDA recommends that the labeling on the package include the primal cut as well as the type of cut. Example; shoulder blade chop or loin chops.

People seem to either like the taste of lamb or to not like it at all. There is not much middle ground. Fresh lamb has a unique almost gamey taste that is somewhat stronger than most fresh meats.

It is the fat on the lamb that gives it much of its flavor. During the cooking process it has an odor that is easily recognized as lamb. Virtually all of the lamb that I have cooked has been outdoors on the grill. My wife (who is not a lamb eater) does not like for me to cook it indoors.

Lamb holds up to stronger seasonings when cooked. Rosemary, garlic, and olive oil are to lamb what jelly is to peanut butter. Mint jelly or a mint jelly sauce is a popular condiment to serve with lamb. I also enjoy a nice jalapeno jelly with lamb.

Most lamb cuts are best enjoyed when cooked to a medium-rare or medium doneness. Rare lamb is not often served. Ground lamb should be cooked to an internal temperature of 160 degrees.

Lamb Leg

Lamb leg is one of the more popular ways to enjoy lamb. Lamb legs are sold in a variety of ways, from bone-in and boneless, to boneless butterflied legs.

Lamb legs are covered with a thin, paper like covering called "fell". The butcher should remove the fell from the leg before offering it for sale.

Bone-in Lamb Leg

Bone-in lamb leg starts with the whole leg, as a "sirloin on" cut. This is a difficult cut for most home cooks to slice and serve. The sirloin bone and the aitch bone in the leg are oddly shaped and hard for most home cooks to navigate. The aitch bone is the hip bone at the end of the femur bone. The sirloin end of the roast is close to 50% bone.

The best way to buy a whole lamb leg is to buy a "sirloin off" lamb leg with the aitch bone removed. Removal of the aitch bone will make the lamb leg easy to carve and serve. The only bone left to navigate around is the femur bone. The femur bone is a simple long round bone.

As you slice the whole leg you will note a pocket of fat in the middle of the roast. There is a fat gland the middle of this fat. Slice down to the fat and remove the fat in one piece before continuing to slice.

Bone-in legs are also sold as shank half and butt half legs. A butt half with the aitch bone removed is an especially good buy because it has little bone to work around. Some butchers remove part of the shank from the shank bone on the shank end to make it a good value as well.

Boneless Lamb Leg

Obviously, the boneless lamb leg is the easiest lamb leg to slice and serve after roasting. The butcher removes the fell from the leg. He then removes the bones and the fat gland. The boneless roast is then shaped into a uniform size and rolled and tied. The boneless roast is also cut in half and sold as a half boneless roast. Boned and rolled, there is no real difference between the butt half and the shank half.

A terrific way to enjoy boneless leg of lamb is to buy a butterflied leg of lamb. The butcher bones the leg in much the same way as he does to make a boned and rolled leg of lamb. The boneless leg of lamb is left untied and the thicker sections are butterflied to an even thickness. (Usually about an inch and a half) The butterflied leg is easy to marinade and grill.

Try a butterflied leg of lamb on the grill for a change of pace during the barbecue season. It is a great way to serve a small crowd. Any leftovers make wonderful sandwiches. All you need is an artisan bread and spicy mustard.

The boneless lamb leg is the perfect cut to make lamb kabobs. You can cut your own or ask your butcher to cut the boneless leg into cubes for you. Cut the boneless lamb into larger cubes an inch and a quarter or more.

Some stores offer a pre-made lamb kabob for sale. As with most ready to cook meat products, you will pay more. Some kabobs also include vegetables. Buying cut vegetables for fresh meat prices seems wrong to me. Cooking the vegetables on kabobs separately is a better way to control the doneness of the meat and the veggies.

Lamb Loin

The lamb loin is the section of lamb that would be T-bone steak if it were beef. The lamb loin is a lean, very tender cut. The prettiest loin chops are those with a large tenderloin section. With or without the tenderloin section, the lamb loin is very tender. You will need at least two 1 ½ inch thick chops for most servings.

Like the rest of the lamb, the loin chops accepts seasonings and marinades well. Loin chops are best cooked over hot dry heat to a doneness of medium-rare to medium doneness. This makes them terrific candidates for a nice hot grill.

Skilled butchers in some specialty meat shops will sometimes bone and roll the lamb loins to make a boneless lamb loin chop, or leave it whole as a roast. In some cases they will add seasoning to the meat before rolling and tying it.

Lamb Rack

The lamb rack is the "prime rib" of the lamb. It has a bit more fat than the loin section. The lamb rack is sold both whole as a roast and sliced into rib chops.

The lamb rib rack roast is the more popular way to enjoy this cut of lamb. The ends of the rib bones are "frenched" before cooking. Frenching is the process of removing the meat and fat between the ends of the bones to make for more fancy presentation when served. A crown roast is made when two or more lamb racks are tied together to form a crown shape.

The most common way to cook a rack of lamb is to pan roast it. The seasoned rack is seared on the outside in a sauté pan and finished cooking in a hot oven for a short time. Sometimes a coating or a glaze is added to the rack before it is finished in the oven.

A whole lamb rack will serve two people or one of my brothers. If I had to, I could probably eat a whole rack as well.

Lamb Shoulder

Lamb shoulder is the chuck meat of the lamb. The shoulder is cut into round bone and lamb shoulder chops. It is also sold as a bone-in and a boneless roast.

Shoulder chops are usually cut less than one inch thick. They are not as tender as loin and rib chops, but they are not a tough cut. Lamb shoulder chops are more than tender enough to pan-fry or grill.

The butcher cuts two to three round bone chops from the shoulder before slicing the rest of the shoulder into blade chops. Some unskilled butchers slice one too many round bone chops from the shoulder or cut them off at the wrong angle. The end result is a round bone chop that has a very large round bone.

Some butchers will package the chops with the large side of the bone hidden on the bottom side of the package. You may think you are buying a chop with a small round bone, when the other side is much larger. Return the chops to the butcher if you do not think you got a fair value.

After cutting off the round bone chops, the butcher slices the shoulder into shoulder blade chops. Shoulder blade chops are a more consistent cut than round bone chops. The remaining neck end of the shoulder is sold as neck meat for stews.

The shoulder with the round bone chops removed is sometimes squared up (the neck end removed) and sold as a bone-in lamb shoulder roast. This is a good cut to ask a customer service butcher to bone and cut into lamb stew meat.

Some butchers remove the bone from the shoulder and sell it as a boneless shoulder roast. The shoulder makes a nice easy-to-serve boneless roast. It is also a good choice to cut into cubes for a traditional Irish stew.

Lamb Shank/Breast

Lamb shank is one of my favorite slow cooked braised meat dishes. The lamb shank is usually the fore shank or front shank. The hind shank cut from the leg of lamb is the premium choice. The hind shank has a better meat-to-bone ratio than the fore shank.

The tougher shank cut has lots of connective tissue and can only be cooked tender by slow cooking. Braising is the perfect way to make these cuts tender. The meat just falls off of the bone when the dish is done.

There are many braised lamb shank recipes available. Most are easy to follow and are well worth the slow cooking time involved. Braised lamb shanks have the same "comfort food" quality as a good pot roast.

The lamb breast includes the small lamb brisket as well as the lamb rib section. The ratio of meat-to-bone is more bone than meat. Braising or slow roasting is the preferred way to cook lamb breast and ribs. Look for recipes on the internet to give this cut a try.

Most often, the lamb breast and ribs are boned out to make ground lamb. It is added to the small amount of trimmings produced when cutting the rest of the lamb, and is used in ground lamb.

Fresh Ground Lamb

Fresh ground lamb is certainly a good way to enjoy lamb at a reasonable price. It can be pan-fried and added to lamb stews or soups.

My favorite way to cook it is to make a grilled lamb patty. It is a nice change from the everyday grilled beef burger. Ground lamb should be cooked to a finished internal temperature of 160 degrees.

Save Money When Buying Lamb

Supermarkets that maintain a selection of fresh lamb for sale every day usually support the sale of lamb by running it as a sub feature in their weekly ad. The best place to buy lamb and to save money when purchasing lamb is to shop in a store that works at selling lamb.

Lamb freezes well and sometimes is even sold frozen. Just follow the basics of packaging any meat product for freezer storage. A nice air tight package is best. Marked down lamb cuts should be frozen the day they are purchased.

Cutting your own lamb stew meat and making your own kabobs are also easy money savers. A sharp knife can be your best money saving tool. Lamb legs are good choice to bone and cut into stew or kabob meat.

Lamb leg is a popular entrée for holiday dinners. Retailers sometimes offer hot deals on fresh lamb legs during the Easter holiday. I have mixed feelings about stocking up during these sales. Normally, I would recommend stocking up at holiday time because often these are the best prices of the year. I would advise caution when buying freezer stock of fresh lamb legs.

In preparation for holiday needs, producers stockpile lamb legs to insure they will have enough. Some of the holiday lamb legs are weeks old by the time they hit the meat case. Even packed in Cryovac®, some of the legs can have a sour odor and a short shelf life when they are opened to process for retail sales.

I am not implying that all of the fresh lamb legs sold during holiday time are past their shelf life. However, I would exercise caution if you are thinking about buying extra fresh lamb legs for the freezer. Lamb legs sold as frozen are usually frozen at their peak freshness. A good price on frozen legs is worth the investment of a couple of extra legs.

GROUND LAMB SLIDERS

My wife, who says that she does not like lamb, loves these sliders. Serve them alone, or as part of a "mixed grill" of sliders. Serve a larger group a variety of ground meat sliders, (ground beef, ground turkey, seasoned ground pork).

1 pound of ground lamb

¼ cup of toasted pine nuts

3 fresh garlic cloves – finely minced, about two tablespoons

Olive oil to coat the patties

Kosher salt and fresh cracked pepper

BASIL MAYONNAISE

This is a terrific condiment for grilled lamb burgers. It is easy to make and will store well in the refrigerator for up to a week. Make it a day ahead so that the flavors will blend.

½ cup of jarred mayonnaise

2 tablespoons of chopped fresh basil

One tablespoon of fresh lemon juice

Toast the pine nuts in a small pan over medium heat. They burn easily, so stay with them until they are lightly toasted. It only takes a couple of minutes. Set aside to cool.

In a small bowl, combine the ground lamb, garlic, and the cooled pine nuts. Take care to not over work the ground meat mixture. It is easy to see when the pine nuts are evenly distributed.

Shape the ground lamb into six patties. Press the center of each patty down so that when they cook and the center swells, the cooked patty will be uniform in thickness. Refrigerate the patties for at least 30 minutes before grilling.

Brush the patties with olive oil and season generously with salt and pepper. Grill the patties on a hot grill to an internal temperature of 160 degrees. To get a nice sear on the meat, allow the patties to cook for five or six minutes before you turn them over. It takes about 15 minutes on an open grill, and is much shorter with the grill cover on. I prefer to cook mine on an open grill. The outside seems to get a crisper crust.

Serve the patties on toasted buns, slathered with basil mayonnaise. Ground lamb is best when served when warm, so no resting period is needed.

JALAPENO GLAZED LAMB RACK

Jalapeno jelly is used to give this lamb rack a sweet spicy glaze. Serve additional pepper jelly on the side for an added kick. There is always something elegant about serving a rack of lamb. One rack (about a pound) will serve two.

Rack of lamb – trimmed of excess fat with the rib bones frenched

Kosher salt

Fresh cracked black pepper

Granulated garlic

2 tablespoons of jalapeno jelly

1 tablespoon of olive oil – plus enough oil to coat the lamb rack

Preheat the oven to 400 degrees.

Coat the rack with oil and season generously with salt, pepper, and garlic. Let the rack rest at room temperature for 30 minutes before cooking.

In an oven safe pan, heat the oil over medium-high heat. Brown the meaty portion of the lamb rack on all sides. It will take about two minutes per side.

With the bone side down, spread the jelly on the meat of the lamb rack.

Roast the lamb for about 15 minutes in the oven. Remove the rack at 125 degrees for medium-rare. The small size of the rack cooks quickly.

Remove the rack from the pan and tent with aluminum foil for five to seven minutes before serving.

Slice the lamb between the ribs to serve.

BUTTERFLIED GRILLED LAMB LEG

The fresh herb and garlic marinade bring out the best in this easy-to-serve lamb cut. Just seeing the lamb marinating in the herb mixture makes my mouth water. Count on three servings per pound for this boneless dish. Leftovers make terrific sandwiches with just a good crusty bread and whole grain mustard.

1 boneless butt half leg of lamb – The butt half is the tender cut of the lamb leg. Ask the butcher to "butterfly" it to an even thickness.

1/3 cup of olive oil

1 tablespoon of finely minced fresh rosemary

1 tablespoon of finely minced fresh thyme

1 tablespoon of finely minced fresh oregano

4 fresh garlic cloves – finely chopped

1 tablespoon of kosher salt

1 tablespoon of fresh cracked pepper

2 tablespoons of red wine vinegar

In a small bowl, mix the oil, herbs, garlic, vinegar, salt, and pepper.

Spread the mixture on all of the outside surfaces of the lamb.

Place the lamb in a glass dish, cover with plastic wrap and refrigerate for four hours to overnight.

Let the lamb sit at room temperature for one hour prior to grilling.

Grill the lamb over indirect heat to an internal temperature of 140 degrees for a medium doneness. Place the lamb on the grill fat-side up.

If you are using a gas grill, turn all of the burners to high and heat for 15 minutes. Turn the burners under the lamb off and close the lid. Check the temperature after about 45 minutes. For charcoal grills, place a pan in the center of the grill and place the coals around the outside for indirect cooking.

Cover the lamb with foil and let it rest for 15 minutes before slicing across the grain in thin slices. Scalloped potatoes or a creamy risotto make an excellent side dish. Serve with mint jelly or better yet, a spicy pepper jelly.

VEAL is most often associated with restaurant cuisine. Fresh veal sales are a very small niche business in most large supermarkets. Veal's higher retails, lower demand and higher shrink make it a break-even product at best. Stores that do sell veal usually support the price with frequent ad retails to help the movement. If you are shopping for the best price in veal, shop where they make an effort to sell it.

Veal is meat from young beef 16 to 18 weeks of age. Male dairy calves provide a significant portion of the commercially raised veal. Dairy cows must give birth to continue producing milk. The veal calves are separated from the cows within three days of birth and are raised in environmentally controlled barns.

The calves are raised in individual stalls. Individual stalls allow the veal farmers to closely monitor the health of each calf. They are fed a milk replacer diet that includes the necessary vitamins and minerals that they require. Some special fed veal calves are fed a diet of nutritionally balanced milk and soy milk products.

Veal farmers use USDA approved antibiotics in the raising of veal to prevent or treat disease. Growth hormones are not approved in the raising of veal.

The production of veal is somewhat of a political football. It is easy to look at the process and to take issue with it. Most of us carnivores have already accepted the life and death process of the foods we like. I grew up in the country and long ago understood where meat comes from. I think that one of the most important lessons young 4-H farmers learn is the sacrifice necessary to provide meat products.

OK, enough of this. Let's get back to the eating of good veal. While most of us eat veal most often in restaurants, cooking veal at home can be just as tasty. Learning to cook restaurant quality dishes at home is always rewarding.

All veal sold in retail stores is inspected for wholesomeness by either the USDA or by state inspectors with equally high standards. Like beef, grading of the meat is voluntary. Most of the graded veal sold in retail stores is Prime or Choice.

The veal is rolled with a grading stamp in the same way as beef. Stores will include the terminology on the label. If the veal is graded Prime or Choice, it will say so on the label.

Like other meats, some veal is sold as a brand name. Look for point of sale pamphlets near the veal display for information about the branded veal.

To extend the selling life many retailers choose to sell veal packaged in modified atmospheric packaging, (MAP). MAP is also an excellent way for brand name veal producers to market their veal. Retailers are able to order only the cuts of veal that sell in their store.

The color of veal ranges from a light reddish color to a very light almost white color. The difference between the colors of veal comes from the feed they are given. The lighter colored veal is often fed a diet of milk. The darker meat is not fed milk. As long as the grade of the veal is high, there is no real difference between the tenderness of the colors of veal.

To me, the most appealing veal to buy is the milk-fed veal. The lighter color is synonymous with higher quality veal. Many chefs and restaurant cooks prefer the darker red meat of the natural veal. I am sure both are good products; however I am still loyal to the early Provimi® brand milk-fed veal I first experienced.

Cooking Veal

Veal should be cooked to an internal temperature of 160 degrees. Tender loin and rib cuts are good cooked over high dry heat such as grilling or pan roasting. Thinner cuts from the round or sirloin are wonderful sautéed with a nice pan sauce. Tougher cuts such as the veal breast or shank benefit from slow roasting or braising. Osso Bucco is a classic dish of slow braised veal shanks.

Veal is a very mild flavored meat. Seasonings and sauces make up a large part of a successful veal dish. If you search for veal recipes you will note that many include an abundance of herbs and additional flavors.

Veal loin

Veal loin is most often sold as a chop. The veal chops look much like smaller versions of T-bone and porterhouse steaks. Choose chops that have a larger tenderloin portion.

One of the easiest ways to enjoy veal chops is to either pan-fry them or to grill them. Marinade the chops in a simple marinade of fresh herbs, olive oil, and the juice and zest of a fresh lemon. Marinate the chops for one to three hours, add fresh cracked black pepper and salt, and grill over high heat.

Veal Sirloin

Veal sirloin is most often sold as a boneless cut. Sliced thin it is good for quick sautés. Sliced thicker, it can be used in much the same way as the veal loin chops.

The veal sirloin is not very large. Sometimes it is sliced thin and included with the veal top round sold as scaloppini. The retails for thin sliced veal top sirloin and thin sliced veal top round are usually very close. The meats are similar in tenderness. I would not pay more for sirloin over the veal top round.

Veal Round

The veal round is mostly boned out and sold as thin sliced steaks for scaloppini. The higher retails for veal make displaying a larger boneless roast a tough sell. If you need a boneless veal roast for a recipe, expect to special order it.

The veal round is separated into top and bottom round sections. In most cases the eye of the round is left on the bottom round. The eye-of-round on the veal is so small that it has no special value on its own.

The veal top round is a preferred cut over the bottom round. Many retailers do not distinguish between the top and bottom round when packaging the veal for sale. Look for slices that are one muscle, with no internal gristle. These will be the top round cuts.

Veal Shoulder

The veal shoulder includes all of the cuts from the veal chuck. The chuck is processed much like the cuts from the lamb shoulder. The chuck is sliced into bone-in shoulder steaks, round bone steaks, and veal stew meat. The chuck is sometimes sold as boneless veal roasts.

Cuts from the veal shoulder are reasonably tender. They can be grilled or sautéed. They are also sturdy enough to hold up to braising and slow roasting.

Veal Shank

The veal shank is most often enjoyed as Osso Bucco. This popular dish is a slow braised thick cut of veal shank cooked in a flavorful sauce. Osso Bucco is a good example of a restaurant dish that is rarely cooked at home.

Cooking Osso Bucco at home can be very rewarding. It makes a nice change-of- pace company dish. You may have to special order veal shanks from your local butcher. Ask for center cut shanks that are at least two inches thick. You can source Osso Bucco recipes in your favorite Italian cookbook or on the internet.

Veal Breast

The veal breast is the brisket of the veal. It is usually sold as a bone-in cut. Cooking methods include slow roasting and braising recipes. Sometimes it is stuffed before roasting. Most meat markets do not regularly stock veal breast. Expect to ask for a special order.

Ground Veal

Ground veal is consumed much differently than most ground meats. First of all it is almost never cooked as a single ingredient. Ground veal on its own has very little flavor. Most ground veal recipes include a variety of herbs, spices and aromatics. The USDA recommends that ground veal be cooked to an internal temperature of 160 degrees.

Ground veal is sautéed and added to spaghetti sauces. Cooked with herbs it is used as a filling for raviolis. Added to ground pork and ground beef, it helps to bind the ground meats together to make wonderful meatballs and meat loaf. Ground veal is also an important ingredient in fresh sausages such as bratwurst.

Most retailers do not have enough demand for ground veal to process ground veal on a daily basis. The other issue is only a small amount of trim is generated when cutting veal.

Ground veal is most often available in MAP or vacuum-packed in one pound packages. This affords the retailer the extended shelf life needed to get a profitable turnover. Follow the shelf life guidelines on the packaged veal. Market ground fresh veal should be used within one to three days of the day of grinding.

VEAL MARSALA

Sweet Marsala wine is the foundation flavor for this classic veal dish. The recipe works well with boneless chicken breast also. Just adjust the cooking time to fully cook the breast before adding it to the sauce. An opened bottle of Marsala wine keeps well in the pantry for several months.

1 pound of veal scaloppini – choose the top round cut if you have a choice. One pound will serve four.

½ pound of sliced mushrooms – I like the brown button mushrooms, but any will do.

1 large shallot or 2 small shallots – thinly sliced.

2 garlic cloves – minced

1 cup of Marsala wine

1 cup of reduced sodium beef stock

Flour – I use a shaker jar with large holes to distribute the flour.

2 tablespoons of unsalted butter

2 to 3 tablespoons of olive oil

Fresh cracked pepper and kosher salt

Season the veal on both sides with salt and pepper. Lightly coat each side with flour.

Add the olive oil to a large skillet and cook the veal over medium-high heat for about one minute per side. Cook in batches and set the cooked veal aside.

Add the butter to the pan and sauté the mushrooms and shallots until lightly browned, four or five minutes. Add additional olive oil if needed. Add the garlic and sauté for one more minute. Shake a couple of tablespoons of flour into the mixture and cook for another couple of minutes.

Add the wine, bring to a boil, reduce the heat to a simmer and cook until the mixture thickens and reduces by about half. Add the beef stock and simmer until sauce thickens again.

Return the meat to the sauce and serve. The dish goes well with buttered egg noodles. Toss cooked noodles with unsalted butter, a bit of freshly grated parmesan cheese, and chopped parsley.

FRESH AND FROZEN POULTRY

I probably should have started my book with fresh poultry. As a nation we consume more fresh and frozen poultry than any other meat. The variety and choices seem to grow daily.

I have always marveled in a country of very diverse lifestyles and opinions that we agree as an overwhelming majority to eat turkey on the same day in November. We agree not only to eat roast turkey but we also agree on most of the side dishes as well.

Chicken is the most consumed meat in America. We eat it hot. We eat it cold. It is on the menu of most restaurants from ethnic, to fine dining and fast food drive- thru. Every family has more than one "go to" recipe for chicken. We eat a lot of chicken!

CHICKEN

Although we have many choices in the way chicken is marketed, most of the chicken raised in America comes from one of two breeds. They are Cornish Hens (a British breed), and White Rock (a breed developed in New England). From these two breeds chicken producers bring to market a huge array of products.

Broiler – Fryer Chicken

Broilers and fryers are the most common size of chicken consumed. Chickens are marketed in the 2 ½ to 4 ½ pound size range. They are sold whole and cut up into all of the different parts we enjoy.

When shopping for whole fryers or chicken parts choose the largest chickens available. It only takes seven to eight weeks to grow a chicken to market size. The bone structure of all of the chickens is a similar size. The meat-to-bone ratio is higher in the larger birds. There is better value in buying larger chickens and chicken parts. The larger chickens are just as tender as the smaller ones.

Many consumers are concerned about the supposed use of growth hormones and antibiotics in the raising of chickens. Given that it only takes a short time to grow a chicken to market size, it seems like a reasonable question.

No hormones are used in the raising of chickens. If antibiotics are used, a withdrawal period is required before the chickens are harvested. This is to insure there are no residues in the fresh chicken. The FSIS (USDA Food Safety Inspection Service) randomly samples birds to test for residue. Their data from these random samples has shown a very low percentage of residue violations.

Some chicken and chicken parts are sold as enhanced product. The chicken is pumped with a solution similar to the water, salt, and sodium phosphate used in enhanced pork. The USDA requires that any additive be clearly stated on the label. Any chicken sold without a statement of additives is just chicken.

All fresh chicken sold in retail stores are inspected by the USDA or state inspectors with equivalent standards. Grading of fresh chickens is voluntary. For the most part only Grade A chickens are sold at retail level.

When shopping for whole fryers or chicken parts choose the largest chickens available. The bone structure of all of the chickens is a similar size. The meat-to-bone ratio is higher in the larger birds. There is better value in buying larger chickens and chicken parts. The larger chickens are just as tender as the smaller ones.

Local and Out-of-State Chicken

One choice many consumers have is to buy chicken that is produced locally or to buy chicken shipped in from another state. The out-of-state chicken is produced by very large chicken producers and it usually less expensive than local chicken. Sometimes there is a difference in the color of the skin of chicken from different sources. The color difference is from the type of feed on which the chicken was raised.

Poultry producers can label their product fresh if it has never been below twenty-six degrees. Any raw poultry stored below zero degrees must be labeled as frozen or previously frozen. Oddly enough, no specific label is required on poultry stored between zero and twenty-five degrees.

Out-of-state producers may take advantage of the lower temperatures to extend the shelf life necessary to get their products to market. Some local producers also "chill" their products prior to shipping. Personally, I have never been able to tell any difference in the taste or quality of local or out-of-state chicken. It is purely an economic choice. It the label does not have a "use" or "freeze-by" date, plan to use the chicken within three days of purchase.

Organic and Free-Range Chicken

Other choices in the fresh poultry case include organic and free-range chicken. Organic chicken usually is also free-range chicken. Free-range chicken is raised in a more open environment with access to the outside. This contrasts sharply with the confined environment of commodity chickens.

Organic chickens are never given antibiotics, and are raised with a diet of organic pesticide-free feed. They follow the same *never ever* guidelines of most organic meat producers. Most organic and free-range chicken displays include POS (point of sale) material nearby. Pick up a brochure to learn the virtues of the organic and free-range chicken in your area.

The cost of free-range and organic chicken is higher than regular chicken. It takes longer and costs more to raise these chickens to market weight. Because of the higher retails the display size and variety may be smaller as well. Some organic and free-range chicken has a better flavor than commodity chicken. The better flavor justifies the higher cost. The best way to tell is to do your own taste test.

Rock Cornish Game Hens

These exotic little hens are just small chickens. They are most often sold as a frozen product. Thaw them in the refrigerator for a couple days. If they are still frosty, finish the thawing process in cold water.

Stuffed and roasted whole, they make a very fancy dinner presentation. Split in half and grilled they still maintain their fancy image.

Roasting Chicken

Roasting chickens sold fresh are the same birds as the fryer/broilers, only much larger. When we used to package poultry in the store, we would pull out the larger fryers to sell as roasters. Just tuck their wings underneath, place in a meat tray, and they were worth more money.

Did you ever notice that when the local supermarket had a sale on whole fryers that the chickens were suddenly much larger than usual?

It the extra week or so that the grower uses to build up the quantity of whole birds for a sale, the chickens get much larger. Sometimes the chickens are five pounds or more. What a great time to buy whole chickens! Remember, the bone structure is roughly the same size, no matter how large the chicken is. The bigger birds are a great value. It is also a great time to get a good value on a large roasting chicken.

Capons

Capons are male chickens up to eight months old that have their reproduction parts surgically removed to encourage larger growth. They weigh up to eight pounds and are thought of as the king of the roasting chickens.

Most capons are sold at Thanksgiving and holiday time as an alternative to a small turkey. Capons are available fresh in some specialty stores and are most often sold frozen in supermarkets.

Because the demand for capons is not year round, ask the butcher if the frozen capons for sale are last year's birds. If so, buy a small turkey or a large fresh roasting chicken.

> For me, the biggest drawback to IQF (individually quick frozen) chicken is that it is almost always an enhanced product. It is pumped with a salt and water solution that adds to the weight of the chicken. Some products also add flavorings to the solution as well.

Stewing Chicken

Stewing chickens are older chickens, sometimes mature laying hens. They can be over a year old. The meat is less tender and is best when cooked with slow moist heat such as stewing.

Stewing chickens used to be a regular part of most poultry displays. Expect to look hard or to special order one if it's on your menu. There is little or no demand for stewing chickens in most supermarket meat markets.

Frozen Chicken Parts

IQF, (individually quick frozen) chicken parts are a popular way to buy chicken in quantity. The IQF chicken parts are sold in the frozen food section in large resealable bags that are easy to close after you take out what you need. The bags are a frequent sale item in most supermarket ads. They are low labor for the store and a good value for the customer.

For me, the biggest drawback to IQF chicken is that it is almost always an enhanced product. It is pumped with a salt and water solution that adds to the weight of the chicken. Some products also add flavorings to the solution as well. If you don't mind the added solution, the IQF chicken can be a good value that is also convenient to use. Personally, I prefer to brine my own fresh poultry if that is what is called for in the recipe.

How to Cut Up a Whole Fryer

Every home cook should know how to cut up a chicken. The whole fryer on ad still represents one of the best values in the meat case. The ability to cook the whole bird, making homemade chicken stock to recipes for all of the parts, is rewarding. Too often chicken served at home means boneless skinless breasts followed by another recipe for boneless skinless chicken breast.

Remove the giblets. Sometimes they are in a package, sometimes they are loose. Reserve the giblets for stock, or fry up the gizzard and hearts if you like them.

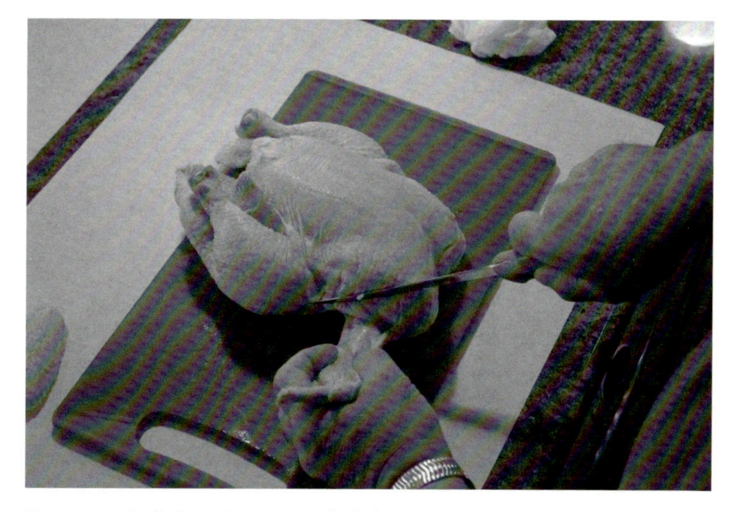

Step one: Pull the wing out to feel the joint connecting the wing. Slice through the corner of the breast then through the wing joint. The wing is a more desirable piece of chicken with a good chunk of the breast attached. Remove the wing tip at the natural joint. Remove the other wing.

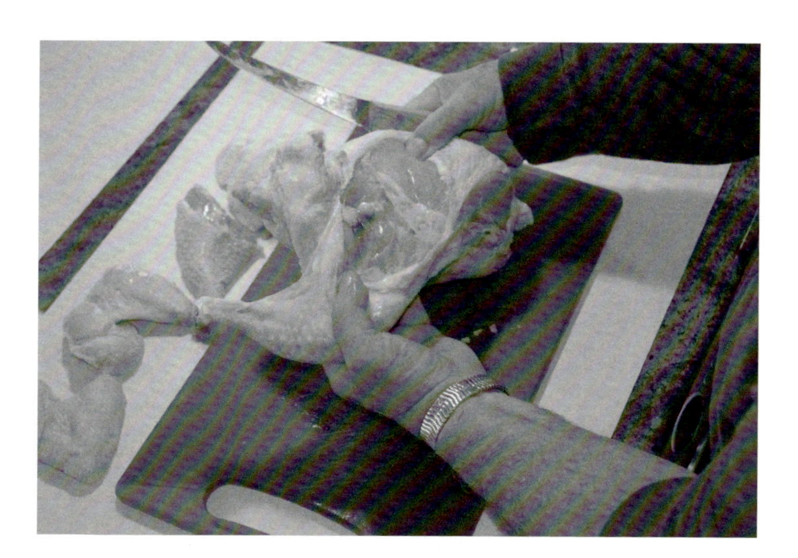

Step two: Cut the skin between the leg and the breast. Grasp the whole leg with your left hand and with your right hand on the body of the chicken bend the leg away and disjoint the leg.

Step three: Holding your knife next to the bone, slice from the back of the bird to the joint that holds on the leg. Push your knife blade into the bird at the joint and pull the leg away sharply. This move will pull out the little pocket of meat on the back still attached to the thigh. Separate the leg from the thigh at the natural joint. Bend the joint and use the knife point to find the soft spot that separates the leg and thigh. Turn the bird around and remove the other leg in the same manner. (Press the blade of the knife against the carcass and pull the leg to remove the pocket of meat with the thigh.)

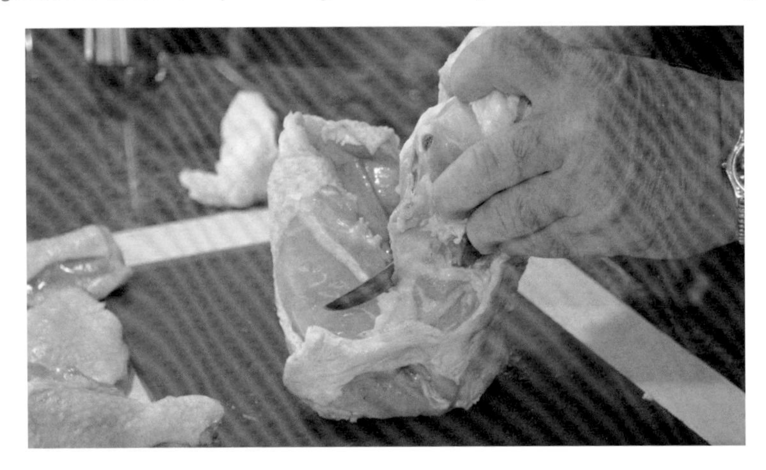

Step four: Turn the bird upside down with the tail in the air toward you. Remove the back from the breast by slicing down firmly to cut through the small rib bones. At the bottom of the cut you will encounter larger bones (including the wishbone) that are hard to slice through. Break the back away from the breast and cut around the larger bones to remove the back from the breast. Set the back aside for use as stock.

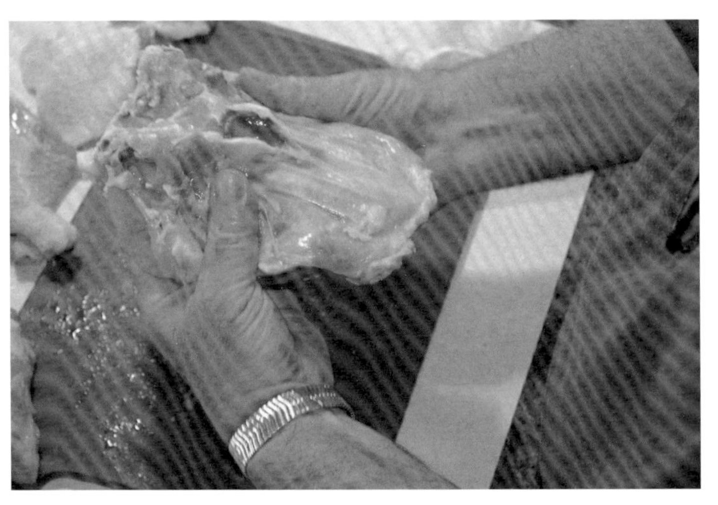

Step five: To cut into bone-in breast, turn the breast skin side down. Using the point of your knife, make a cut through the white cartilage at the thicker end of the breast. Bend the sides of the breast away from you to pop the keel bone loose. With the keel bone removed you can now slice the breast into two pieces.

Step six: Cut the breast into two sides. Trim any excess cartilage, skin or fat. The rib bones are easy to remove if you prefer boneless breast. Freeze the back and neck to make stock at a later date.

Saving Money When Shopping for Chicken

The obvious way to save money on your chicken purchase is to shop the weekly ad in your local supermarket. Fresh chicken is a regular feature in most meat marketing plans. The popularity of chicken and its benefit to the retailer as a lower loss ad item keep it in the ad on a regular basis.

Because of its relatively low price point, retailers can offer chicken at a hotter price than higher priced meats. The total sales for a chicken ad are a lower percentage of the total weekly sales.

The ad writer starts each weekly ad plan with an average markup of the sales at regular price, (without anything on ad). The average markup for the weekly sales goes down as prices and margins are lowered in the ad plan.

These negative dollars are called **ad loss dollars.** Sales of lower priced ad product, like chicken, generate a lower total sales dollar amount. This lower dollar amount equals a lower percentage of the total weekly sales. The end result is a weekly ad with a planned higher profit margin.

After choosing chicken on sale, the next best way to save money is to choose larger chickens or chicken parts. Because of the short time that it takes to bring a chicken to market, the bone structure of all of the birds is similar in size and weight. Larger birds have a better meat-to-bone ratio. It's simple, larger chickens have more meat.

Learning how to cut up a whole fryer is a good way to save money as well. The boneless skinless breast that is the most popular chicken part is also the most costly part of the bird. Learning to enjoy the rest of the chicken is an easy money saver.

Unless the prices are the cheapest you have ever seen, I don't recommend stocking your freezer too full of sale chicken. Although it freezes well, chicken is on sale often enough to eat it fresh most of the time.

If you do freeze extra chicken, thaw it in the refrigerator. If you need to thaw it more quickly, use a sink full of cold water that you change every 30 minutes. The two to three days it takes to thaw frozen chicken for use is one of the reason I rarely freeze chicken.

ROAST CHICKEN WITH LEMON

This roast chicken makes a wonderful Sunday dinner. Roast two birds and use the extra meat for easy midweek tacos, salads, or sandwiches. Lemon zest on the outside, and cut up lemons in the cavity give this roast chicken its fresh flavor. The lemons on the inside help to keep the meat moist during roasting.

One whole fryer roaster – 4 ½ to 5 pounds

2 tablespoons of olive oil

2 small fresh lemons or one large lemon

Kosher salt, fresh cracked black pepper, and granulated garlic

OPTIONAL

4 Yukon gold potatoes, skin left on and quartered or

8 baby red potatoes cut in half.

1 Medium onion – peeled and quartered

Preheat the oven to 400 degrees. Remove the giblets and rinse the inside of the cavity of the chicken. Pat the bird dry with paper towels.

Rub the outside of the bird with the olive oil. Generously season the chicken on all sides with salt, pepper, and garlic. Tuck the wings underneath the whole chicken.

Spray a roasting pan with nonstick spray. I use a 9 by 14 inch glass casserole dish about two inches deep. (A pan deep enough dish to collect the juices during roasting)

If you are adding the optional veggies, layer them in the bottom of the pan. Place the seasoned chicken breast side up on the veggies or in the empty pan.

Zest one of the lemons across the top of the chicken. Quarter both lemons and stuff them in the cavity of the bird.

Roast for one hour and 15 minutes or until the internal temperature in the breast reaches 170 degrees. Baste every 15 minutes for the last 30 minutes of roasting. Baste one last time when you remove the chicken from the oven. Slide a large knife into the cavity of the bird to help support it when you move the chicken to a serving plate. Allow chicken to rest for 10 to 15 minutes before serving.

Return the veggies to the oven to brown while the chicken rests. For a nice sauce, strain the cooking liquid. Melt 2 tablespoons of butter in a sauce pan. Add 2 tablespoons of flour. Cook the flour and butter for two to three minutes, until it starts to brown slightly. Add the pan liquid and heat over medium heat until thickened.

ROAST CHICKEN FRENCH DIP SANDWICH

Purchase a pre-roasted chicken or roast an extra chicken when you roast one. A little bit of chicken goes along way. The recipe is a great way to get more than one meal from a roast chicken. Tender moist chicken, wonderful gruyere cheese, tarragon mayo on a toasted roll dipped in toasted shallot chicken stock……Mmmmm.

1 ½ pounds of sliced pre-cooked roast chicken – Serves four, a generous six ounces per serving.

4 soft sandwich rolls

1 tablespoon unsalted butter – for sautéing the shallots

1 box of low sodium chicken broth

1 large or 2 small shallots – sliced thin

1 tablespoon of olive oil

½ pound of gruyere cheese – grated or sliced thin, substitute any good Swiss cheese

Kosher salt and fresh cracked pepper

Tarragon mayonnaise – recipe follows

TARRAGON MAYONNAISE

2 tablespoons of fresh tarragon leaves – coarsely chopped

2/3 cup of mayonnaise

½ fresh lemon juiced

1 tablespoon of olive oil

Use a deep saucepan to sauté the shallots in olive oil and butter over medium heat. When the shallots start to brown and caramelize, add the chicken broth. Leave the pan on the heat and reduce the liquid by one-third to one-half. Season the broth with salt and pepper to taste.

To make the tarragon mayonnaise, add the tarragon, mayonnaise, lemon juice, and olive oil to a small bowl and mix well.

Cut the rolls in half and toast under a broiler. Spread the tarragon mayo on both sides of the rolls.

Add the chicken to the bottom of the roll. Sprinkle the cheese on top. Melt the cheese under the broiler. It doesn't take very long, so keep a close eye on them. Cover with the tops and cut in half on a diagonal. Serve the sandwiches with a small cup of the broth for dipping.

CHICKEN AND CHEESE STUFFED POBLANO PEPPERS

This is another great recipe to use leftover roasted chicken. Poblano peppers are a mild, flavorful pepper. Together with the cheese and chicken mixture they make an easy change-of-pace dinner.

1 ½ cups of shredded roast chicken – shred the chicken as you pull it from the bone. Discard the skin and gristle.

4 medium-sized poblano peppers

3 tablespoons of red onion – diced fine

¼ cup of sour cream

½ cup of queso fresco, shredded – look for it in the Mexican food section of your supermarket

1 tablespoon of smoked paprika

½ teaspoon of cumin

Small can of corn, 8 ounces – drained

2 tablespoons of chopped cilantro

In a large bowl combine the chicken, sour cream, cheese, onion, paprika, cumin, corn and cilantro until well-mixed.

Wash the peppers. Place the peppers large side down on cutting board. Remove the top quarter or so of the pepper. You want an opening large enough to stuff the pepper. Using a knife and a spoon, remove the seeds and membranes. Rinse the loose seeds out with cold water.

Stuff the peppers with the chicken mixture.

Place the peppers in a 9 by 12 glass casserole pan that has been sprayed with a nonstick cooking spray. Use the pepper tops (skin side down) to help level the stuffed peppers. Grate a bit of queso fresco over the chicken mixture.

Roast in a preheated 350 degree oven for 30 minutes.

Grate more queso fresco over the peppers before serving.

Serve with Mexican-style brown and wild rice with a dollop of Mexican crema. (Mexican sour cream) Add one tablespoon of fresh lime juice to ¼ cup of sour cream as a substitute.

MEXICAN-STYLE BROWN AND WILD RICE

I use a premixed blend of brown and wild rice with soft wheat and rye berries. You can substitute any mix of brown and wild rice. The rice mix has a nice sweet nutty flavor.

1 cup of brown and wild rice mix

1 large yellow onion – diced

2 slices of thick cut pepper bacon – cut into ¼ inch dice

2 tablespoons of olive oil

14.5 ounce of diced tomatoes – look for the fire roasted tomatoes if they are available

2 cups of low sodium chicken stock

1 tablespoon of dried oregano

Using a deep saucepan over medium heat, sauté the onions, bacon, and the olive oil. Sauté until the onions are translucent and the bacon starts to crisp.

Add the rice and toss to coat with the oil for one minute. Add the chicken stock, diced tomatoes, and oregano. Stir well and bring to a low boil. Reduce the heat to a simmer and cover the pot with a lid.

Simmer the rice for 30 minutes. Check the rice two or three times during the last 10 minutes to make sure it has enough liquid to finish cooking. Add another ½ cup of chicken stock if needed.

Serve with a dollop of Mexican crema. (Mexican sour cream) If you cannot find Mexican crema, add a couple of tablespoons of fresh lime juice to one cup of sour cream.

CHICKEN-IN-A-POT

The name for this wine braised chicken comes from one of my favorite Rickie Lee Jones songs, "Danny's All-Star Joint". "Chicken-in-a-pot" is verse Rickie sings in her unique sexy soulful style. Serving this dish in my home always results in a group rendition of "Chicken-in-a-pot, chicken-in-a-pot, chicken-in-a-pot."

8 large fryer thighs – Skin removed and trimmed of any visible fat.

4 tablespoons of olive oil – One to rub on the thighs for the seasonings to stick and three to brown the meat.

Kosher salt, fresh cracked black pepper

Flour to coat the chicken – Tip: Keep flour for coating meat in a cheese shaker container with large holes. It uses less flour and isn't wasteful like dredging meat in a container of flour.

1 tablespoon of unsalted butter

2 shallots – Sliced thin

1 cup of mushrooms – I use rehydrated morels because my family keeps me well supplied from their mushroom hunting. Any kind of fresh or rehydrated mushroom works well. Cut small mushrooms in half and slice larger ones thick.

2 garlic cloves- sliced thin

2 cups of red wine – Any full bodied red wine that you like to drink.

4 sprigs of fresh thyme

Preheat the oven to 350 degrees.

Rub the thighs with olive oil and season generously with salt and pepper. Coat the thighs in flour. Shake off any excess flour.

Heat the remaining oil in a deep cast iron pot over medium-high heat. Brown the chicken, working in batches so as not to overcrowd the pan. Remove all of the chicken to a large plate.

Reduce the heat to medium. Add the butter and sauté the mushrooms and the shallots for five to seven minutes. Add the garlic and sauté for one more minute. Add the wine. Return the thighs and any juices to the pan. Toss the shallots, mushrooms, and wine to coat the chicken. Add the thyme.

Roast covered for 30 minutes. Remove the pan from the oven. Take off the top layer of chicken and rotate it with the chicken on the bottom. Return to the oven, uncovered, for 30 more minutes.

Take the thighs out and put them in a large serving bowl. Remove the thyme sprigs from the sauce. Pour the sauce over the thighs and serve. Just a nice salad and some good crusty bread for sopping sauce is a delightful dinner.

TARRAGON CHICKEN/APPLE SALAD WITH PINE NUTS

This is a refreshing chicken salad that is even better if made a few hours or a day ahead. Serve with flakey croissant rolls or another light roll for a late lunch or dinner. This recipe will serve four generously.

1 ½ pounds of boneless skinless fryer breast – Season with salt and pepper and grill or pan-fry. Allow to cool before using.

5 crisp, sweet apples – I like Fuji or Honey Crisp – Remove the core and cut into one inch cubes with the skin on.

1 fresh lemon – Juiced

2 stalks of fresh celery – Remove the fibrous ribs with a peeler and dice ¼ inch.

3 tablespoons of fresh tarragon – Remove the leaves from the stem and chop

¼ cup of toasted pine nuts – Toast until lightly brown over medium heat. They will burn quickly, so stay with them.

¾ cup of mayonnaise

Be careful to not overcook the fryer breast. It is done just as soon as it is firm to the touch. Boneless fryer breast will quickly become dry if over cooked.

Dice the chicken breast into one inch cubes.

Dice the apples and add the lemon juice to prevent browning

Mix all of the ingredients together in a larger bowl.

Move the salad to a nice serving bowl. Cover with plastic wrap and refrigerate for at least an hour. *The salad can be made one day ahead. Like many dishes, the salad is better served the next day.*

Allow the salad to warm to room temperature for 15 to 20 minutes before serving.

Leftover salad will keep for two to three days in the refrigerator

BUTTERMILK DIPPED FRIED CHICKEN

This fried chicken is a take on the fried chicken my Mom used to make. In the summertime she would make a big batch with potato salad. When my Dad came home from work we would take a watermelon to a nearby lake for a mid-week picnic dinner. The chicken is fall-apart tender from the time spent in the low temperature oven. Eight pieces will serve four.

8 large fryer thighs – Remove the skin and any visible fat. I use thighs because they are my favorite part. Taking the skin off is my attempt at a healthy lifestyle. You can substitute any chicken parts, skin on if you like. *This is not a crispy fried chicken recipe. The time in the oven makes the coating soft.*

¾ cup of low trans-fat vegetable oil

Kosher salt, fresh cracked black pepper, onion powder, and granulated garlic

1 cup of flour – Seasoned with one tablespoon of kosher salt, one teaspoon of fresh cracked black pepper, one teaspoon of granulated garlic, one teaspoon of onion powder, and ½ teaspoon of cayenne pepper (optional). Mix the seasonings together and then stir in the flour to get a more even mix.

1 cup of low fat buttermilk

Preheat the oven to 300 degrees. Place a wire rack on a baking sheet and spray with non-stick spray. This will allow the excess fat to drain away.

Season the chicken generously with salt and pepper.

Set up a dredging station. Put the buttermilk in a bowl large enough to dip the chicken. You need another bowl for the flour, and a baking sheet or large plate to hold the breaded chicken.

Lightly coat the seasoned chicken with flour.

Dip the chicken in the buttermilk, shake off the excess.

Dip the chicken in the seasoned flour, shake off the excess.

Heat the oil in a deep 12 inch fry pan over medium-high heat.

Brown the chicken for three to four minutes per side, until it is a nice light brown color. Place the pieces of browned chicken on the wire rack on a baking sheet. Place the baking sheet full of chicken in the oven and roast for 30 minutes.

Serve the fried chicken warm or cold with my mother-in-law's famous potato salad or a fresh green salad and soft rolls.

MARY'S FAMOUS POTATO SALAD

Every family seems to have someone who makes the best potato salad. My Mother- in-law, Mary's, potato salad has been the standard for years. This is my version of her salad.

2 pounds of Yukon gold potatoes – Choose potatoes of like size so that they all cook the same. Peel the potatoes and cut them in half.

1 tablespoon of apple cider vinegar

1 medium size red onion

3 to 4 celery stalks

4 large hard boiled eggs – Cut into ¼ inch or less dice

¾ cups of mayonnaise

1 tablespoon of celery seed

1 teaspoons Kosher salt, ½ teaspoon of fresh cracked black pepper

Bring the potatoes to a boil in a pot of salted water with the tablespoon of vinegar. Reduce the heat to a simmer. Start checking the potatoes for doneness after 12 minutes. The vinegar will help to keep the potatoes from becoming mushy.

Drain the potatoes in a large colander. They will continue to cook and the surface of the potato will dry as they cool.

Peel the onion and dice fine. Rinse the cut onion under cold water in a fine sieve. This helps to take away the overpowering taste.

Using a peeler, remove the fibrous ribs from the celery. Cut the celery lengthwise into slender sticks. Slice the sticks into a fine dice.

In a large bowl, mix the celery, onion, eggs, celery seed, salt, pepper, and mayonnaise together.

By now the potatoes should have cooled to a point that they are easy to work with. Dice the potatoes into a ½ inch dice and fold into the rest of the salad. Taste, and adjust the salt and pepper if necessary.

Remove the salad to a smaller serving bowl. Cover and refrigerate overnight before serving

FRESH AND FROZEN TURKEYS

Turkey is so much more than a favorite holiday roast turkey dinner. Today's consumer includes turkey in the menu on a regular basis. The poultry section of most supermarkets includes a variety of turkey parts, ground turkey, and fresh turkey sausages. Fresh ground turkey is a staple of many families as a healthy alternative to ground beef. The smoked section also includes a variety of smoked turkey sausages, turkey bacon, and smoked turkey parts and ham products.

Turkey Production Standards

All turkeys sold in retail stores are inspected by the USDA or state inspectors with equivalent standards. Like chickens, the grading stamp is voluntary. Grade A is the highest grade and the only grade you are likely to see at retail level.

When I first started in the meat business we used to offer Grade C and Grade B turkeys as a lower cost choice at thanksgiving time. These lower grade birds usually had tears in the skin or had a wing missing. With today's demand for turkey, these lower grade turkeys are used in the production of value-added turkey products.

The standards for raising turkeys to market are very similar to the standards for raising chickens. No hormones are used in the raising of turkeys. If antibiotics are used a withdrawal period is required to insure there are no residues are left in the meat at harvest time. Like chickens, the USDA Food Safety Inspection Service (FSIS), randomly samples for antibiotic residue. Any raw poultry shown to contain residues above the tolerance levels is considered to be adulterated and cannot be sold.

Whole Holiday Turkey

The roasting of the Thanksgiving turkey can be intimidating to even the most seasoned home cook. This once a year ritual can be hard to duplicate even if last year's turkey was the best ever. I don't think it is any coincidence that all of the traditional side dishes are moist. We have all eaten more dry turkey than we care to remember.

My Mother-in-law, Mary, and I have been cooking the Thanksgiving turkey for a number of years now. We take pride in serving a moist tender turkey.

Every year Mary takes notes of what temperature we used and the time it took to roast the turkey. Another piece of the puzzle is that I always buy the same size turkey. Keeping notes and using a similar size turkey helps us to better duplicate last year's success.

Helpful Hints to Roasting a Whole Turkey

Use a meat thermometer. If your turkey comes with a pop-up plastic timer, consider that it is just a plastic pop-up timer and use a real thermometer.

Use a meat thermometer. A whole turkey is food safe when cooked to an internal temperature of 165 degrees. Place an oven safe thermometer in the thickest part of the breast or in the thigh, without touching the bone.

Tuck the wing tips under the bird. Season the outside of the bird and the cavity with salt and pepper. Place the whole bird in a shallow roasting pan.

Add cut apples, onions, citrus, and fresh herbs to the cavity before roasting. Add any combination that sounds good to you. The added moisture from the inside helps to cook the bird and to keep it most.

Add ½ cup of water to the bottom of the pan.

Cover the pan with a lid, or tent with aluminum for the first 1 ½ hours. Tuck the aluminum in around the pan, under the level of the top to insure any moisture drips back into the pan.

Remove the cover to brown the turkey. Replace the cover until the bird comes up to temperature. Baste the bird during the browning process.

Allow the turkey to rest for 20 minutes before carving. Hint: remove each side of the breast as a whole piece to slice. It is much easier than taking slices off of the whole breast.

Roast the turkey at 325 to 350 degrees. Roast the turkey unstuffed. The USDA recommends that the stuffing be cooked separately to an internal temperature of 165 degrees.

Allow three to four hours for a 12 to 14 pound bird and four to five hours for a 20 to 24 pound bird. Keep notes of the time and temperature each year to duplicate your success.

Fresh Turkey

Most consumers would look at a new shipment of fresh turkeys received in a meat market and call them frozen. Fresh turkeys can be "chilled" down to a temperature of twenty-six degrees and still be labeled "fresh". Trust me, a turkey shipped at twenty-six degrees feels frozen.

With the production time needed to meet the huge demand for holiday dinners, most fresh turkeys are chilled down to twenty-six degrees. Obviously, chilled turkeys will have an extended shelf life. This extra storage and shipping time is necessary to build up the quantities needed.

Handing a customer a rock-hard fresh turkey, special ordered two weeks early at a premium price, is a tough sell for most butchers. "Old school butchers" left the pallets of fresh turkeys out of refrigeration in the back room to soften before the customers picked them up.

Responsible retailers in today's food safe world encourage customers to pick up their holiday turkey a couple of days early and give them directions on how to "thaw' their fresh turkey.

Because they are not frozen down to zero degrees like turkeys labeled as frozen, they soften up more quickly than frozen birds. A couple of days in the refrigerator followed by a short time in cool water are all that is necessary to soften a chilled fresh bird.

Fresh "chilled' turkeys usually thaw to a fresh softness in the meat case display in a day or so. They may appear to be fresher than the still hard special order turkey, but they are all usually the same batch.

Frozen Turkey

Frozen turkeys are held at zero degrees or below. They must be labeled frozen or previously frozen when sold. Oddly enough, no specific labeling is required for poultry between zero and twenty-six degrees.

It takes significantly longer to thaw a frozen turkey than a chilled fresh turkey. It can take up to six days to thaw a 24 pound turkey in the refrigerator. Allow twenty-four hours for every four to five pounds of frozen turkey to thaw in the refrigerator. It is food safe to refreeze a turkey that has been properly thawed in the refrigerator.

Thawing a frozen turkey in cold water is also a food safe way. Leave the protective wrap on the turkey. Allow 30 minutes per pound and change the water every 30 minutes. A 24 pound turkey can take twelve hours, so you might want to do it in a couple of stages. Refrigerate the turkey between stages. A turkey thawed in water is food safe but should not be refrozen. If you want to refreeze the turkey, thaw it in the refrigerator. It takes longer, but the temperature is a constant 40 degrees or less.

The choice of whether to buy a fresh or a frozen turkey is a question of price and convenience. Fresh turkeys cost more but are more convenient to get ready to roast.

Frozen turkeys are usually sold at cost or below as the holiday sales driver. A cheaper retail price makes them well worth any extra trouble. The opportunity to buy any product at a fraction of its wholesale cost is hard to ignore. Retailers frequently sell turkeys costing 75 cents or more for retails of 25 cents or less.

I would pass on most early holiday season turkey sales. The first turkeys offered for sale each Thanksgiving season may be last year's crop of frozen turkeys. Many holiday ads run for multiple weeks. I always wait until later in the ad to buy my bird to make sure I get this year's production.

Hen or Tom?

Many older people are convinced that the larger tom turkeys are tougher. At one time, this was true. With today's modern production methods, the only difference between a hen and a tom turkey is the size. It takes about the same time to raise both. The toms just get bigger.

The best answer I ever heard to a customer's question about the difference between hens and toms came from a young apprentice meat cutter. His reply to the customer's question was "Madam, toms come two to a box and hens come four to a box."

Basted Turkey

Some brand name turkey producers differentiate their product by selling "basted" turkeys. To add moisture to the bird they add a solution containing butter or other edible fat, seasoned water, or other spices and flavor enhancers. The injected solution cannot add more than three percent to the total weight. The ingredients must be listed in descending order of amount.

Like other enhanced meats, the addition of extra moisture helps to keep the final cooked product moist. It is still very possible to overcook a basted bird to a dry doneness.

My personal choice is still a non-basted Grade A turkey. If you want to add moisture to your turkey before roasting, try brining.

The basic recipe for brine is 1 ½ cups of Kosher salt to one gallon of water. If you use table salt, choose non iodized salt and reduce the amount to one cup per gallon. Add more flavors with the addition of sugar, peppercorns, and dried berries or herbs.

Organic and Free-Range Turkeys

Producers of organic and free-range turkeys most often sell their product as a brand name turkey. Establishing a brand name allows them to build a loyal following to their product. POS (point of sale material), allows them opportunity to tout the virtues of their product.

Producers of organic turkeys raise them in much the same way as other organic meat producers. They are raised on a diet of organic feed with no use of antibiotics. The USDA does not allow the use of growth hormones in the raising of any turkeys, including non organic and free-range turkey. Most organic turkeys are also raised in the "free-range" method.

To be sold as "free-range", turkeys must be allowed access to the outside. The turkeys are allowed to roam outside of their coop. Some feel that this added exercise adds a more natural flavor to the harvested bird. There is also the argument that it is a more humane method of raising turkeys.

My personal experience with free-range turkey is limited. My first and only free- range Thanksgiving turkey was not a positive experience. There was little or no fat on the bird. As it cooked, the aroma of Thanksgiving in the oven was missing. Although it was cooked to the proper temperature, the meat was dry and less flavorful. The next year we went back to a less expensive "smell good" turkey.

I am sure there are many people who prefer the free-range turkey. I respect their choice. My personal choice is still the less expensive non-basted Grade A bird. Grade A is as good as it gets, and the lower cost is the deciding factor.

Fresh Turkey Parts

Most supermarkets carry a good selection of precut turkey parts. Much of the precut turkey parts are displayed in MAP (Modified Atmospheric Packaging). The extended shelf life of MAP allows the retailer to offer a larger selection and the consumer a longer time to store the product at home. This win, win combination has dramatically increased the sale of fresh turkey parts over the last few years.

Some retailers also process fresh turkey parts in-store. They use the turkey part business to offer fresh whole turkeys for sale on a daily basis. Fresh turkeys are cut into parts before their shelf life expires. The whole birds are replaced by new stock with fresher dates.

Thawing and cutting frozen turkeys into parts enables the retailer to move through excess frozen inventory. As part of the effort to make sure they do not run out of holiday turkeys, stores are frequently left with an overstock of frozen turkey.

Turkey parts cut from frozen turkeys should be labeled as "previously frozen". Likewise, basted frozen turkeys cut into parts should be labeled with an ingredient list of the basting ingredients. Unfortunately, I don't think this happens everywhere. It does not impact the wholesomeness of the product. However, it is information to which the consumer is entitled.

The consumer demand for whole frozen turkey after the holidays is very low. Retailers cannot afford to offer pricing similar to holiday ads. Competition for holiday sales is very fierce. Stores build a loss into their yearly budgets for the holiday sales period.

Stores frequently offer turkeys priced well below the actual cost. A turkey offered for sale at 25 cents per pound probably cost the retailer 75 cents per pound or more. An after holiday ad of 79 cents per pound just does not look like a good deal to the consumer.

Frozen turkeys are never cheaper than during Thanksgiving sales. Buy extra turkeys for future Holiday and family dinners.

Saving Money Shopping for Turkey

It should be pretty obvious that the best time to save money when shopping for turkey is the Thanksgiving sales period. Buying at prices 70% below wholesale is a no brainer. If you serve turkey for Christmas, buy it at Thanksgiving. Buy an extra smaller bird to serve next summer at a family gathering.

Buying extra frozen turkeys to thaw and cut into parts is also a good idea. It is OK to refreeze the turkey after cutting it into parts as long as you keep it chilled and refreeze it right away. Thaw turkeys to be refrozen in the refrigerator.

Cutting a whole turkey into parts is not much different than cutting up a whole chicken. The hardest part is removing the metal or plastic clip that trusses the legs together.

Step one: With the clip out of the way, remove the giblets. Sometimes they are in a package, sometimes they are loose. Reserve the giblets for stock or fry up the gizzard and hearts if you like them.

Step two: Turn the bird upside down with the wings toward you. Pull the wing out to feel the joint connecting the wing. Slice through the joint to remove the wing. Remove the wing tip at the natural joint. Slice through the center joint of the wing to cut it into two pieces. Turn the bird over and remove the other wing and cut it into pieces the same way.

Step Three: Turn the bird right-side up with the legs facing you. Cut the skin between the leg and the breast. Holding your knife next to the bone, slice from the back of the bird to the joint that holds the leg on. Grasp the whole leg with your left hand and with your right hand on the body of the bird, bend the leg away and disjoint the leg.

Push your knife blade into the bird at the joint and pull the leg away sharply. This move will pull out the little pocket of meat on the back still attached to the thigh. Separate the leg from the thigh at the natural joint. Bend the joint and use the knife point to find the soft spot that separates the leg and thigh. Turn the bird around and repeat the process.

Step four: Place the breast skin-side up to process into boneless breast. Slide your knife along the keel bone to remove each side. The turkey breast tenderloins will separate easily from the boneless breast. They are usually large enough to use in a separate dish. Pounded thin between sheets of plastic wrap, they make wonderful breaded turkey cutlets.

The remaining boneless breast can be used in a variety of ways. It can be roasted, skin on. You can remove the skin and slice it into easy sauté turkey slices. It can be sliced thin and used in stir-fry dishes.

Step five: The remaining carcass is perfect for making stock and soups. Break the back section into two easier to manage pieces. Season and roast the parts in a 350 degree oven for 45 minutes to add color and flavor to your stock.

Basic turkey stock recipe – Add turkey carcass, 2 onions quartered, 3 chopped carrots, 3 chopped celery stocks, 1 head of garlic cut in half, and 2 bay leaf to 6 to 8 quarts of water. Bring to a boil, and simmer for 3 hours. Strain the liquid and pick the meat from the carcass to add to a soup. The stock can be reduced further for a more flavorful stock. Freeze in containers for later use.

The butcher's basic rule of thumb for cutting a whole turkey into parts is: the retail value of the breast equals the retail value of the whole bird. You can sell the breast for the same total retail value as the whole bird. The sale of the legs, thighs, and wings is just gravy.

Using this logic you would be money ahead to cut up your own turkey at regular prices. Buying turkeys at 70% of wholesale cost makes it a steal.

A twenty pound turkey at 25 cents per pound will cost you $5.00. Serving a family of four you will yield six to eight or more meals with a meat cost of less than 20 cents per serving. Here are some of the meals you could get from one bird.

- Oven roasted turkey drumsticks – served with the traditional trimmings
- Braised turkey thighs – use as a substitute for a veal Osso Bucco recipe
- Roast skin on boneless turkey breast – roast with a compound herb butter under the skin
- Sliced skinless turkey breast cutlets – a quick and easy mid week sauté dinner
- Turkey breast tenderloin cutlets – pounded thin, rolled in seasoned flour and pan-fried. Serve with pan gravy and mashed potatoes.
- Turkey wings – Jumbo hot wings. Season and slow roast the wings for one hour in a 350 degree oven. Toss with your favorite hot sauce. Serve with celery sticks and cool blue cheese dressing.
- Turkey noodle soup – Sauté 2 diced carrots, 1 diced onion, and 2 diced celery stocks until tender. Add turkey meat, 1 teaspoon of dried thyme and 5 cups of stock. Simmer for 30 minutes. Add frozen egg noodles and cook until tender.

Ground Poultry

The room given to fresh ground poultry in the ground meat section is a good indicator of its growing popularity. Ground turkey is a popular option to lean ground beef for many families. Most large supermarkets offer a selection of light and dark turkey as well as ground chicken.

The USDA guidelines for ground poultry are different than the guidelines for ground beef. There is no standard for regulating the amount of fat that ground poultry may contain.

Most of the fat in ground poultry comes from the skin. When making ground poultry, the ratio between meat and skin can be no more than natural proportions. Because the skin is 10 to 15 percent by weight, the amount of fat is self-regulating. Producers are not allowed to add additional skin or fat to ground poultry products.

Some ground poultry producers market their product with a stated fat content. Others market their product as skinless. Another way to market ground poultry is to identify the cut that it is ground from (such as ground turkey breast).

Ground poultry producers are not required to identify the cuts of poultry used to make the grind. Ground turkey breast meat is marketed as just cuts from the breast. Most ground turkey is sold as dark turkey or light turkey.

Ground poultry can be substituted for ground beef in most recipes. Keep in mind that ground poultry needs to be cooked to an internal temperature of 165 degrees to be food safe. This compares to a final temp of 160 degrees with ground beef.

Cured and Smoked Poultry Products

The USDA reserves the name "ham" for only products from the back leg of the hog. Turkey that is cured and smoked to resemble ham is not allowed to be called ham. After the marketing name, the product made from dark turkey meat is always followed by the statement "cured turkey thigh meat".

The amount of retail display space allotted to smoked and cured poultry seems to get larger every day. Virtually any smoked and cured pork or beef product is available in a turkey or chicken version.

Smoked and cured poultry products are sometimes priced lower than similar beef and pork products. There is also a perceived health benefit from eating lower fat poultry products.

It pays to read the labels when comparing. I know of one frozen pork sausage product that lists the number one ingredient as turkey skin. Unlike unseasoned ground poultry products where the skin content is limited to natural proportions, manufacturers can include whatever ingredients are in their recipe. All they have to do is list them in descending order of most to least.

Some of my favorite poultry products include fresh poultry sausages. Poultry sausage makers seem to be more creative in adding seasoning to their sausage. They use everything from fresh herbs and cheese to sundried tomatoes and mangos. The mild ground poultry seems to mix well with a variety of seasonings.

Ground poultry can be used as a base to make your own fresh sausage. Follow the same recipe used for pork sausage the first time you make it. There are pork sausage recipes in the pork section of this book.

TURKEY THIGH OSSO BUCCO

This wonderful slow-braised dish is more commonly made with veal or lamb shanks. Turkey thighs are the most flavorful part of the turkey. When braised, they cook to mouth watering tenderness.

2 turkey thighs – remove any visible fat

Flour – enough to coat the turkey

3 tablespoons of olive oil

Kosher salt, fresh cracked black pepper, granulated garlic

½ red onion – diced

2 medium carrots – peeled and cut into bite size pieces

1 cup of button mushrooms – stemmed and quartered

1 can of diced tomatoes – 15 ounce can, choose fire roasted if available

1 cup of dry white wine – it is OK to substitute chicken stock

1 small bunch of fresh thyme

Preheat the oven to 350 degrees

Season the turkey generously with salt, pepper, and garlic. Using a cheese shaker, sprinkle enough flour on the turkey to coat.

Heat 2 tablespoons of oil in a heavy cast iron pan over medium heat. Brown the turkey thighs on both sides. Remove the turkey to a large plate.

Add the onion, mushrooms, and carrot to the pan. Add the extra olive oil if needed. Sauté until the onion is translucent.

Add the tomatoes and wine to the pan. Deglaze the bottom of the pan. Return the turkey thighs and any juices to the pan. Toss in the sprigs of fresh thyme.

Cover and roast for one hour in the oven. After one hour, remove the cover and continue to cook for another 45 minutes (until the sauce reduces and the turkey is fork tender).

Discard the thyme, and serve the turkey and sauce on a large platter. Buttered egg noodles with a bit of chopped parsley go well with this dish.

HOWEVER you enjoy it, meat is a basic food group to us carnivores. I hope this book guides you to eating and enjoying your fair share.

GLOSSARY

Ad loss – The dollars lost from regular price when selling meat on sale. Ad loss dollars are usually represented as a percent of the total profit dollars.

Aitch bone – The hip bone on the back leg of the meat. In most cases it is removed to make cutting and cooking easier.

Aromatics – Herbs and fresh vegetables that add a flavorful aroma to the finished dish

Boxed Beef – The commodity name for beef processed into subprimals and sold to retailers by the box.

Butterfly – A cutting technique where the meat is not completely sliced through and is then opened up to make a larger, thinner cut.

Cap meat – A thicker solid muscle found on the outside of cuts like top round, top sirloin, and beef ribs. Cap meat is usually less tender and removing it enhances the remaining cut.

Charcuterie – Centuries old methods of curing and smoking meats.

Chine – The bottom edge of the backbone on the rib cut. The butcher uses the meat saw to remove the chine bone. With the chine bone removed the rib can be cut with a knife between the ribs.

Cryovac® –Vacuum-packaged meat. Thick plastic bags are used to vacuum-package subprimals for boxed beef.

Cuber – The machine used for tenderizing meats such as cube steak. The meat is run between two rollers with sharp cutting blades to tenderize the meat.

Deglazing – The process of adding liquid to a pan to remove the fond for making sauces.

Enhanced meat – Meat that has been injected with a salt or flavoring solution. Enhanced meat is also called pumped meat.

Fell – The thin paper like covering on fresh lamb. Most often, it is removed when the fat is trimmed. If not, it should be trimmed before cooking.

Fond – The browned bits of flavor left in the pan after sautéing. Deglazing loosens the fond for making sauces.

Freezer burn – Dry areas of frozen meat caused by holes in the packaging or freezing meat for too long. Minor freezer burn can be trimmed off before cooking. Excessive freezer burn can add an "off" flavor to the cooked meat.

Frenched – The meat cutting technique for exposing the rib bones by removing the meat between the rib bones. French cutting the rib bones is most often done on lamb or pork rib roasts.

Grind log – The meat market's tool for tracking production of ground meats. Information usually includes date and amount of grind, fat content, and source of product ground. The grind log is a valuable tool for tracking the history of ground meats in a recall.

Gristle – Larger, thicker strips of connective tissue or collagen. Slow moist cooking methods turn the gristle to gelatin. External gristle should be removed from tender dry cooked cuts.

IQF – Individually quick frozen

Kevlar® – A synthetic cut resistant material used to make cutting gloves. Kevlar is a registered trademark of DuPont.

MAP packaging – Modified atmosphere packaging. The oxygen in the sealed package is replaced with another gas such as nitrogen. MAP packaging extends the shelf life of the product.

Margin dollars – The profit dollars before shrink and operating expenses. Margin dollars are usually noted as a percentage of total sales.

Meat grain – The direction the fibers run in the meat. Meat is more tender when sliced across the grain.

Membrane – The thin covering on the inside of the rib bones.

Oxymyoglobin – The process where the myoglobin in the meat tissue reacts with oxygen to turn the meat a bright red color.

PSMO – Peeled, side muscle on, is the meat industry name for whole peeled beef tenderloins. PSMO is the boxed beef term.

POS- Point of sale material. Usually brochures placed near the product to tout its virtues.

Purge – The liquid that accumulates in Cryovaced® packages of meat during the wet aging process. Also, the natural juices that leak out of raw meat.

Shelf life – The suggested time you can keep meat refrigerated, not frozen.

Shrink – The dollar loss in product lost in the sales process. Shrink is usually noted as a percentage of total sales dollars.

Silverskin – The thin shiny membrane on the outside of some cuts such as beef and pork tenderloin. Remove before cooking.

Tare – The weight of the package and packaging material. The tare is subtracted from the total weight of the package.

Yield – The saleable meat after the cutting loss, discarded fat, bone, and purge.